VOLUME EIGHTY EIGHT

# ADVANCES IN
# COMPUTERS

## Green and Sustainable Computing: Part II

VOLUME EIGHTY EIGHT

# Advances in
# COMPUTERS

## Green and Sustainable Computing: Part II

Edited by

**ALI HURSON**

*Department of Computer Science*
*Missouri University of Science and Technology*
*325 Computer Science Building*
*Rolla, MO 65409-0350*
*USA*
*Email: hurson@mst.edu*

Amsterdam • Boston • Heidelberg • London
New York • Oxford • Paris • San Diego
San Francisco • Singapore • Sydney • Tokyo
Academic Press is an imprint of Elsevier

Academic Press is an imprint of Elsevier
225 Wyman Street, Waltham, MA 02451, USA
525 B Street, Suite 1900, San Diego, CA 92101-4495, USA
The Boulevard, Langford Lane, Kidlington, Oxford, OX51GB, UK
32, Jamestown Road, London NW1 7BY, UK
Radarweg 29, PO Box 211, 1000 AE Amsterdam, The Netherlands

First edition 2013

**Library of Congress Cataloging-in-Publication Data**
A catalog record for this book is available from the Library of Congress

**British Library Cataloguing-in-Publication Data**
A catalogue record for this book is available from the British Library

ISBN: 978-0-12-407725-6
ISSN: 0065-2458

For information on all Academic Press publications
visit our web site at *store.elsevier.com*

Printed and bound by CPI Group (UK) Ltd, Croydon, CR0 4YY

13 14 15   10 9 8 7 6 5 4 3 2 1

Working together to grow
libraries in developing countries
www.elsevier.com | www.bookaid.org | www.sabre.org

ELSEVIER    BOOK AID International    Sabre Foundation

# CONTENTS

# PREFACE

Traditionally, Advances in Computers, the oldest Series to chronicle the rapid evolution of computing, annually publishes several volumes, each typically comprising of five to eight chapters, describing new developments in the theory and applications of computing. The theme of this 88th volume is similar to the 87th volume: "*Green and Sustainable Computing.*" It is a collection of five chapters that covers a diverse aspect of related issues.

Green computing, as defined by Sam Murugesan in 2010, refers to the "study and practice of designing, manufacturing, and using computer hardware, software, and communication systems efficiently and effectively with no or minimal impact on the environment." Adopting the aforementioned definition, this volume covers a wide range of solutions ranging from hardware to software. The aim is to inform the reader of the state of the art and science of green and sustainable computing. The chapters that comprise this volume were solicited from authorities in the field, each of whom brings to bear a unique perspective on the topic.

In Chapter 1, "Energy-Aware High Performance Computing—A Survey," Knobloch articulates the issue of power management within the scope of high performance computing and its future. As noted in his contribution, to build an Exascale supercomputer (expected by 2020), one has to take a holistic approach to the power efficiency of data centers, the hardware components, and the software support system. The chapter provides an excellent tutorial and concludes with a discussion about "The eeClust Project," a joint project between several universities and research centers funded by the German Ministry of Education.

In Chapter 2, "Micro-Fluidic Cooling for Stacked 3D-ICs: Fundamentals, Modeling, and Design," Shi and Srivastava are looking deep into the proper cooling techniques for three dimensional integrated circuit. The work claims that the conventional air cooling mechanisms might not be enough for cooling stacked 3D-ICs where several layers of electronic components are vertically stacked. Several aspects of 3D-ICs with micro-channel heat sinks are investigated, the existing thermal and hydrodynamic modeling and optimization for 3D-IC with micro-channels are discussed, and a new micro-channel based runtime thermal management approach is introduced. The proposed approach dynamically controls the 3D-IC temperature by controlling the fluid flow rate through micro-channels.

The concept of wireless communication at the chip level is the main theme of Chapter 3. In this chapter Murray, Lu, Pande, and Shirazi argue that multi-hop communication is a major source to the performance limitation of massive multi-core chip processors, where the data transfer between two distant cores can cause high latency, power consumption, and higher temperature, which ultimately decreases reliability and performance, and increases cooling costs. Consequently, the chapter articulates the so-called "small-world on-chip network architecture," where closely spaced cores will communicate through traditional metal wires, but long distance communications will be predominantly through high-bandwidth single-hop long-range wireless links.

In Chapter 4, "Smart Grid Considerations: Energy Efficiency vs. Security," Berl, Niedermeier, and de Meer bring out the efficiency and stability of the current power grid in coping with volatile renewable power production. In their view the "Smart Grid" is a cyber-physical system (CPS), and consequently, shares many challenges that other cyber-physical systems are facing. This chapter studies the current status and future developments of the Smart Grid and its challenges. It covers enhancements in terms of energy efficiency and new energy management approaches. Finally, the discussion analyzes some of the threats concerning the new Smart Grid infrastructure and addresses the interdependencies between energy efficiency and security in the Smart Grid.

Finally, in Chapter 5, "Energy Efficiency Optimization of Application Software," Grosskop and Visser detail energy efficiency at the high-level application software. In this article they propose a shift in perspective to the design of energy-efficient systems: the high level application software. They articulate two reasons for such a shift: (1) the application is the primary unit of evaluation and design for energy-efficiency, and (2) the need to raise the level of analysis from the hardware, data center infrastructure or, low-level system software, to the high-level application software. Extension of the Wirth's law, "dramatic improvements in hardware performance are often canceled out by ever more inefficient software," is the main motivation for this discussion.

I hope that you find these articles of interest, and useful in your teaching, research, and other professional activities. I welcome feedback on the volume, and suggestions for topics for future volumes.

Ali R. Hurson
Missouri University of Science and Technology,
Rolla, MO, USA

CHAPTER ONE

# Energy-Aware High Performance Computing—A Survey

## Michael Knobloch
Jülich Supercomputing Centre, Forschungszentrum Jülich GmbH, 52425 Jülich, Germany

## Contents

*Advances in Computers*, Volume 88
ISSN 0065-2458, http://dx.doi.org/10.1016/B978-0-12-407725-6.00001-0

## Abstract

Power consumption of hardware and energy-efficiency of software have become major topics in High Performance Computing in the last couple of years.

To reach the goal of 20 MW for an Exascale system, a holistic approach is needed—the efficiency of the data center itself, the hardware components, and the software have to be taken into account and optimized.

We present the current state of hardware power management and sketch the next generation of hardware components. Furthermore, special HPC architectures with a strong focus on energy-efficiency are presented.

Software efficiency is essential on all levels from cluster management over system software to the applications running on the system. Solutions to increase the efficiency are presented on all that levels, we discuss vendor tools for cluster management, tools and run-time systems to increase the efficiency of parallel applications, and show algorithmic improvements.

Finally we present the eeClust project, a project that aims to reduce the energy consumption of HPC clusters by an integrated approach of application analysis, hardware management, and monitoring.

# 1. INTRODUCTION

## 1.1 What is HPC?

High Performance Computing (HPC), or supercomputing, is basically the solution of very difficult computing intensive problems in a reasonable time with the help of the fastest computers available. Such problems arise in various fields of science and industry, making supercomputing an essential tool in those areas.

In science, simulation is now considered the third pillar besides theory and experiment as it helps scientists to tackle problems that would be unsolvable without supercomputers. For example, in physics, where the scales became either too small (e.g. particle physics) or too large (e.g. astrophysics) to be able to experiment further, supercomputing is the only way to test hypotheses that could lead to new discoveries. In addition, in many other fields of science HPC became an indispensable tool, be it biology, (bio-)chemistry, material sciences, mathematics and computer science, or weather and climate research. HPC is no longer only a scientific tool. Model cars or planes have been simulated for cost saving in developing fewer number of expensive prototypes at early design and implementation process. Similarly, simulations are becoming a norm in many other disciplines such as oil and gas exploration, medical imaging, and financial markets, to name a few.

HPC began in the 1960s with the first machines designed by Seymour Cray, the pioneer of supercomputing. Cray dominated the HPC market in the 1970s and 1980s, beginning with the Cray 1 in 1976, a vector processor-based supercomputer [136]. Later Cray improved the vector processing further, but stuck to a relatively low processor count. In the 1990s massively parallel supercomputers like the Cray T3E [6] and the Intel Paragon [18] began to appear which changed the programming model from vector processing to message passing. This led to the development of MPI [122], the message passing interface, which is still the dominant programming model in HPC.

All supercomputers at that time where specifically designed HPC machines but commercially not successful. That made supercomputers very expensive, and only a few centers could afford them. This situation changed with the upcoming of cluster computers, i.e. the connection of several inexpensive commodity off-the-shelf servers via a (high-speed) interconnect. Those machines are commonly known as Beowulf clusters after the first machine build in 1994 by Thomas Sterling and Donald Becker at NASA [150]. Today the HPC market is divided in custom massively parallel supercomputers and commodity clusters with or without special accelerators like GPGPUs (General Purpose Graphic Processing Units).

Parallel Programming is inherent with HPC and is becoming a standard in software development with the upcoming of multi-core processors. The typical programming models used in HPC are message passing with MPI for inter-node communication and threading models like OpenMP [21] or POSIX threads (Pthreads) [33] on shared memory nodes. New programming paradigms like PGAS (Partitioned Global Address Space) languages, where the whole memory of all nodes is considered as one contiguous virtual memory with implicit communication, are on the rise as they ease programming by removing the explicit expression of communication [164]. However, data locality matters for such programming model and has to be ensured by the application developer to get any reasonable performance.

### 1.1.1 Who has the Fastest—The Top500 List

With more and more machines installed the competition between the HPC centers for the fastest machine began to rise. So a method to rank supercomputers had to be found.

The first Top500[1] list was created in 1993 by Hans Meuer and Eric Strohmaier. It ranked the systems according to the Flops (Floating-Point Operations per Second) rate achieved in the High Performance Linpack (HPL) benchmark by Jack Dongarra et al. [38]. Since then the list has been updated twice a year at two of the major HPC conferences, the ACM/IEEE Supercomputing Conference (SC)[2] in November in the USA and the International Supercomputing Conference (ISC)[3] in June in Germany.

Figure 1 outlines the historical and projected performance development of the systems in the Top500 list. If the projections hold we can expect an Exaflop system at around 2019.

The latest Top500 list of June 2012 is dominated by the IBM Blue Gene/Q (see Section 2.3 for more details on that architecture) with 4 systems in the top 10, the largest of which is the Sequoia supercomputer at Lawrence Livermore National Lab with more than 16 Petaflops (=$10^{15}$ floating-point operations per second) sustained performance in the Linpack benchmark, replacing the K computer (from the Japanese word *kei*—meaning 10 quadrillion), the first 10 Petaflops machine at RIKEN Advanced Institute for Computational Science campus in Kobe, Japan as the #1 system in the world.

### 1.1.2 Is One Number Really Enough?

However, using just the Linpack result as a metric to describe supercomputers led to lots of discussions within the HPC community as it prefers certain types of machines and the workload which is not representative of real-world applications.

Lots of other benchmarks were developed like the HPC challenge benchmark suite [109], the NAS parallel benchmarks [12], or the SPEC benchmark

[1] www.top500.org
[2] www.supercomputing.org/
[3] http://www.isc-events.com

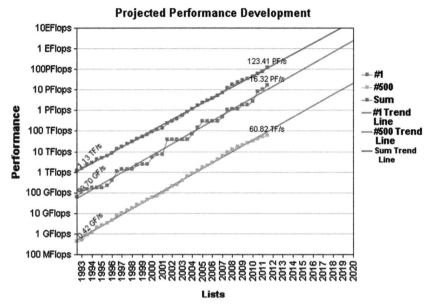

**Fig. 1.** Top500 projected performance development for the fastest and the slowest system in the list. From the extrapolation of the data an Exaflops system is expected in the 2019/20 time frame. Source: www.top500.org.

suites for MPI[4] and OpenMP[5] that better reflect real HPC applications. All consist of several kernels that are representative for a certain type of application.

Although those benchmark suites are widely used as reference to evaluate new tools and methods, none of them could replace the HPL as no other benchmark could match the beauty of the HPL of producing just one number for ranking the systems.

## 1.2 HPC Goes Green

For a long time, performance was the only design constraint for supercomputers, no one really cared about the power consumption of the machines or the corresponding environmental impacts. So the machines became more and more powerful and more and more power hungry. Figure 2 shows the development of the sustained performance (Fig. 2a) and the power consumption (Fig. 2b) of the machines in the Top500 list for the #1 system and the average of the top 10 and the Top500.

[4] http://www.spec.org/mpi2007/
[5] http://www.spec.org/omp/

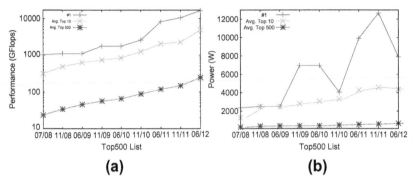

**Fig. 2.** Development of performance (a) and power consumption (b) in the Top500 list.

However, the power capacity of HPC centers is becoming a problem as feeding several Megawatt of power is not trivial and requires a lot of infrastructure to support, which is reflected in the operational costs of the center. Each Megawatt of power consumption increases the electricity costs alone by approximately 1 million USD yearly (at 12 ct/kWh). In Europe, where the price for electricity is significantly higher, that problem is even more severe.

In addition to the high cost, the environmental impact became an issue as well. Data centers have a significant share in the global $CO_2$ emission. Many governments promised to reduce their countries carbon footprint, and as many HPC centers are publicly funded they have to prove an efficient operation to guarantee the continuation of their funding. Wu-chung Feng from Virgina Tech was among the first to identify power as an issue for HPC and he suggested to list that ranks not the most powerful but the most efficient supercomputers [144].

The Green500[6] list [28] started at the SC'07 to complement the performance oriented Top500 list and re-rank the supercomputers on that list by their Flops/W efficiency in the Linpack benchmark. Figure 3 shows the development of the efficiency of the #1 system and the average of the top 10 and all 500 systems in the list. While there is a huge improvement in the efficiency of the top systems, the overall efficiency did not increase much.

On the Green500 list of June 2012, the top 21 spots are all held by Blue Gene/Q systems with an efficiency of over 2.1 GFlops/W, a huge margin of more than 700 MFlops/W to the first non Blue Gene/Q system, an Intel test system with Xeon processors and MIC accelerators ranked #22 with

---

[6] www.green500.org

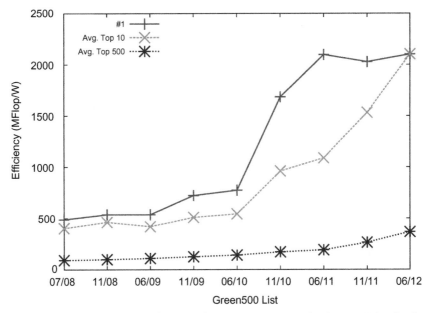

**Fig. 3.** Development of the efficiency of supercomputers in the Green500 list for the number 1 and the average of the top 10 and Top500. In November 2010 the first Blue Gene/Q prototype entered the list which led to the huge increase in efficiency for the #1 system at that time. In June 2012, the top 10 systems are all Blue Gene/Q based with the same efficiency. However, the overall efficiency of the Top500 did not increase equally.

1380 MFlops/W efficiency. The Blue Gene/Q marks a continuation of the success Blue Gene/P line which held the top spots in the first two Green500 lists.

In general high–density custom supercomputers lead the list, followed by accelerator-based clusters and normal commodity clusters [142].

The DARPA, the American Defense Advanced Research Projects Agency, commissioned a large study to determine the challenges on the race to Exascale [17]. The study led by Peter Kogge defined four major challenges for creating an Exaflops machine, which are energy and power, memory and storage, concurrency and locality, and resiliency. The authors claim that an Exaflops system, a machine that will be able to perform $10^{18}$ floating–point operations per second, should consume at most 20 MW of power—which corresponds to 50 GFlops/W. To reach this goal the power efficiency needs to be increased by a factor of 25 compared to today's most efficient system, the IBM Blue Gene/Q.

This is only possible by a holistic approach to reduce the power consumption of a HPC data center—power efficiency must be increased at all levels from the data center itself over hardware down to the software.

As energy-efficiency became an emergent topic several general and well-established conferences organized main tracks for energy-efficiency along with corresponding workshops. Even dedicated conferences like the International Conference on Energy-Aware High Performance Computing (EnA-HPC[7]) were established and dedicated books have been published [62]. A community of researchers working in Green-HPC was created, e.g. the Energy Efficient High Performance Computing Working Group (EEHPCWG[8]). At the Supercomputing Conference 2011 a new portal for Energy Efficient HPC[9] was launched. The SciDAC dedicated their fourth Workshop on HPC Best Practices to power management in HPC centers [34].

Nevertheless, HPC centers still need to invest in "brainware" for Green-HPC [20], i.e. hire staff those are able to apply green methods in the center and they need to create an awareness among users how much energy their applications actually consume.

## 1.3 The Quest for the Right Metric

Green-HPC is about power consumption and energy-efficiency. Power is the amount of electricity some hardware consumes at a specific moment in time while energy is the integration of power consumption over time.

### 1.3.1 Power vs. Energy

Distinguishing between power and energy is useful as there are significant differences when considering metrics for data centers, hardware, and software. Both power and energy are important for Green-HPC, yet in different contexts. Power is more important from a data center and hardware perspective as a data center can be fed only with a certain amount of power. From an application perspective, energy is the more important factor, i.e. the power consumption over time.

But both have to be regarded together. It makes no sense to run the most efficient application in a data center where most of the power is wasted for cooling the equipment, and inefficient software can waste an enormous

---

[7] www.ena-hpc.org
[8] http://eehpcwg.lbl.gov/
[9] http://www.eehpc.com/

amount of energy even on low-power hardware. For both data centers and applications several metrics to evaluate the power- and energy-efficiency exist.

### 1.3.2 Data Center Metrics

The green Grid,[10] a non-profit, open industry consortium with the aim to improve the resource efficiency of data centers and business computing ecosystems, formulated two metrics to determine the efficiency of a data center: PUE and DCiE. Intel provides a white paper on how to measure the PUE and improve the DCiE [117].

The Power Usage Effectiveness (PUE) is the fraction of the power spend for the IT equipment ($P_{IT}$) on the total power of the facility ($P_{TOT}$):

$$PUE = \frac{P_{TOT}}{P_{IT}}. \tag{1}$$

As the total facility power is composed of the power spend for IT equipment and the power spend for the remaining infrastructure, $P_{TOT} = P_{Infrastructure} + P_{IT}$, Eqn (1) can be formulated as

$$PUE = \frac{P_{TOT}}{P_{IT}} = 1 + \frac{P_{Infrastructure}}{P_{IT}}. \tag{2}$$

So the lower the PUE of a data center the better with a minimum of 1.

The Data Center infrastructure Efficiency (DCiE) is the reciprocal of the PUE defined as:

$$DCiE = \frac{1}{PUE} = \frac{P_{IT}}{P_{TOT}}. \tag{3}$$

Typical data centers have a PUE between 2.0 and 1.7 and good centers reach a PUE of about 1.4.

The University of Illinois' National Petascale Computing Facility[11] aims for a PUE of 1.2 or even better, achieved by a number of measures that could serve as a model for future data centers. Water is cooled by air in three on-site cooling towers for free, and power conversion losses are reduced by running 480 V AC power directly to the compute systems. Further the machine room is operated at the upper end of the possible temperatures defined by the ASHRAE (American Society of Heating, Refrigerating, and Air-Conditioning Engineers) which reduces cooling effort.

[10] http://www.thegreengrid.org/
[11] http://www.ncsa.ill inois.edu/AboutUs/Facilities/npcf.html

However, the major disadvantage of PUE and DCiE is that they make no statement of the the computational efficiency as they do not define how much of the power spend for IT equipment is actually spent to do useful work (though it is difficult to define useful work).

### 1.3.3 Application-Level Metrics

On the application side it is much harder to find a suitable metric to express the energy-efficiency of the application. The metric used in traditional HPC is just the time-to-solution, i.e. the performance of the application without any focus on the energy consumption. A corresponding green metric would be the energy-to-solution metric without any focus on performance. Both are of course not suitable metrics for Green-HPC, the first would not be green and the latter not HPC. Feng et al. [73] discussed possible metrics and ended up with the metric they use for the Green500 list, the Flops/W metric.

A commonly used metric is the energy-delay-product (EDP), i.e. the energy consumed by an application multiplied by the runtime which considers both energy consumption and performance. Bekas and Curioni proposed a new metric [16], the $f(time\text{-}to\text{-}solution) \cdot energy$ (FTTSE) which is basically a generalization of the EDP. They discuss several possible functions $f(t)$ and their implications on the metric. This metric is certainly not generally applicable but currently the best we have.

## 1.4 Outline of This Chapter

In this chapter we will discuss many aspects of power and energy-aware high performance computing from both hardware and software perspectives. Section 2 is dedicated to the power management features of hardware. We discuss currently available components in Section 2.1 and give an outlook into future developments in Section 2.2. Further, energy-aware HPC architectures are presented in Section 2.3. In Section 3 we discuss the influence of software on the energy-efficiency of a system. We present various tools from hardware vendors (Section 3.1) and and research groups covering schedulers (Section 3.2), power measurement and modeling in Section 3.3, and tools to optimize the energy consumption of HPC applications in Section 3.4. Additionally, we give an overview of energy-aware applications and algorithms in Section 3.5. The eeClust project is presented in Section 4. It includes an application analysis, which is discussed in Section 4.2, a novel hardware power-state management (Section 4.3), and the eeMark benchmark for computational performance and efficiency, which is presented in Section 4.4. Finally, we conclude this chapter in Section 5.

## 2. HARDWARE POWER MANAGEMENT

Power consumption is a hot topic in enterprise data centers for quite some time now and even longer so within the domain of mobile computing where reduced power consumption directly translates in longer battery runtime. This motivated the hardware vendors to add more and more power management features to their devices to reduce the overall energy consumption. In this section we will give an overview of the hardware power management features that are available for server CPUs, memory, interconnects, and storage. Moreover, we discuss which of those features can be efficiently exploited in the context of HPC. Further we outline current developments in these areas which might have a significant influence on the power consumption of a HPC-system. Finally we present some special HPC-architectures developed with a strong emphasis on energy-efficiency.

Figure 4 shows a typical power distribution within a (HPC-)server, although of course the real values vary from system to system. About 50% of the power (excluding power supply losses) is consumed by the processor, 20% by the memory subsystem, 20% by the interconnect and storage subsystem, and 10% for the rest likes fans, etc. So the CPU is the primary target for power management, and it is the component with the most sophisticated power management features. However, memory and network power consumption are not to be underestimated especially for large-scale installations and the SoC-based architectures discussed later.

Desirable in the long run would be energy-proportional hardware as expressed by Barroso and Hölzle, both from Google. They made a case for energy-proportional computing in [13] discussing the situation at Google,

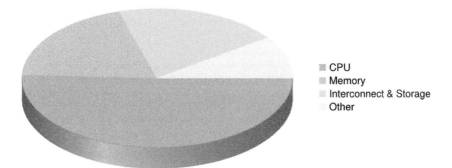

**Fig. 4.** Typical power distribution in a server.

but that can be transferred to most other data centers. Energy-proportional computing means that the power consumption of a component or system scales with its utilization. When a normal server is idle it still consumes about 50% of absolute power and at 50% utilization—a typical utilization in enterprise data centers—it is already at over 75% of peak power, yielding in a very poor energy-efficiency below 70–80% utilization. Tolia et al. gave an overview of how energy-proportionality can be increased even with non-energy-proportional hardware by optimizing the whole machine setup and interaction of components [155]. They use live migration of virtual machines to get a better utilization of single servers of the system and shut down the unused systems. While this approach might work in enterprise data centers, it is not suitable for a HPC center. Minartz et al. proposed an approach to calculate the potential energy savings for HPC application using power-manageable hardware [115]. The simulation framework they present uses application trace files with hardware utilization information to determine an upper bound for possible power savings using energy-proportional hardware. They were further able to simulate different types of energy-aware hardware to compare the possible energy savings. This method might also be used to determine the most efficient system for an application if there is the choice between several systems (as discussed in Section 3.2).

## 2.1 State-of-the-Art

In 1996 Intel, Microsoft, and Toshiba developed an open standard for device power management by the operating system, the Advanced Configuration and Power Interface (ACPI) [67], to consolidate and replace the wide variety of existing proprietary power management solutions. ACPI provides platform independent interfaces for hardware discovery, configuration, power management, and monitoring and became the common power management interface for many components. Several global states and device states were defined by the ACPI interface:

- Global power states (ACPI G-States)
  The ACPI specification defines four global power states with six sleep states:
    - G0/S0 (Working)—the system is up and running.
    - G1 (Sleeping)—this state divides into four sleep states S1 to S4:
      * S1—the processor caches are flushed and no further instructions are issued. Processor and Memory are still powered, but all unnecessary devices may be powered down.

> \* S2—the CPU is powered down, the dirty cache is flushed to memory.
>
> \* S3 (Suspend to RAM)—the current state of the system is written to memory which remains powered. The rest of the system is powered down. A restart loads the contents from memory and resumes at the saved state.
>
> \* S4 (Suspend to disk)—the content of the main memory is saved on disk, and the system is powered down. A restart loads the data from disk and resumes at the suspended state.

- G2/S5 (Soft Off)—The system is shut down without saving the current state. A full reboot is necessary to return to G0. Some devices might still be powered to enable for example Wake-on-LAN.
- G3 (Mechanical Off)—The system's power is completely off, i.e. the power supply unit does not distribute power any more. Typically only the real-time clock is still running, powered by its own battery.

- Device power states (ACPI D-States)
  Every ACPI compliant device must implement the following four D-States:
  - D0 (Fully on)—the device is fully operational.
  - D1 and D2—intermediate power states. These states may vary from device to device.
  - D4 (Off)—the device is turned off and does not respond to its bus any more.

- Processor idle sleep states (ACPI C-States)
  CPU idle states, the so-called C-States, are defined more precisely than for other devices. C-States are states where the CPU has reduced or turned off selected functions. Generally, higher C-States turn off more parts of the CPU, which significantly reduces the power consumption. Different processors support different numbers of C-States in which various parts of the CPU are turned off. Processors may have deeper C-States that are not exposed to the operating system. However, four C-States have to be implemented (analogous to the D-States):
  - C0 (Active)—the processor is fully operational.
  - C1 (Halt)—the processor does not execute any instructions and can return to C0 with very little delay. This C-state is required for any ACPI-compliant processor, but sub-states are possible like the C1E (Enhanced Halt State) on some Intel processors.
  - C2 (Stop-Clock)—the processor turns off the core clock allowing it to shut down logical units. Further the L1 cache may be flushed to

save power. Returning to C0 takes significantly longer than returning from C1. This is an optional state.

- C3 (Sleep)—cache coherence is not maintained any more, meaning the L2 cache can be flushed or turned off. This is also an optional state. Some processors also have variations of the C3-State (Deeper Sleep-States). The deeper the Sleep-State, the longer it takes for the processor to return to the operational state.

- Power states (ACPI P-States)
  While in operational mode (C0) a device can be in one of several power states (P-States). P-States provide a way to scale the frequency and voltage at which the processor runs so as to reduce the power consumption of the CPU. The number of available P-States can be different for each model of CPU, even those from the same family. Generally P0 is the state where the CPU runs at maximum frequency and voltage:
  - P0—max. frequency and voltage.
  - P1—frequency and voltage lower as P0.
  - ...
  - Pn—frequency and voltage lower as P(n − 1).

### 2.1.1 CPUs

The CPU is not only the main power consumer in classic HPC servers (see Fig. 4), but also the component with the most power management features.

### The Power Consumption of a CPU

The power consumption $P$ of a microprocessor is composed of a static and a dynamic fraction:

$$P = P_{\text{dynamic}} + P_{\text{static}}. \tag{4}$$

The dynamic power can be roughly expressed as a function of the supplied voltage $V$ and the processor frequency $f$:

$$P_{\text{dynamic}}(V,f) = A \cdot C \cdot V^2 \cdot f, \tag{5}$$

where $A$ is the activity on the gates and $C$ is the capacitance of the processor (which depends on the size of the die and the number of transistors).

The static part of the processors power consumption depends on the leakage current $I_{\text{leakage}}$ and the supplied voltage and is given as:

$$P_{\text{static}} = I_{\text{leakage}} \cdot V^2. \tag{6}$$

For current microprocessors, the dynamic fraction of the power consumption dominates the static one, but with decreasing production processes the influence of the leakage current, and thus the static fraction, rises.

For any given processor the capacitance is a fixed value, but frequency and voltage might vary, although not independently from each other [123]:

$$f_{max} \propto (V - V_{threshold})^2 / V. \tag{7}$$

Equation (7) shows that the maximal frequency is basically proportional to the applied voltage of the processor so when the frequency is reduced the voltage can be reduced as well, which can lead to a significant reduction of the power consumption according to Eqn (5).

Basically all modern server processors support Dynamic Voltage and Frequency Scaling (DVFS) to reduce the voltage and frequency of the processor in several steps. That corresponds to the ACPI P-States, which strongly depend on the processor and might even vary for different processors in the same family. Switching P-States is usually quite fast, making it the primary target to exploit in HPC where the aim is to minimize run-time dilation. The transition between P-States is controlled by the operating system via so-called governors and can be done either automatically in idle times (with the ondemand governor) or directly by the user (with the userspace governor), enabling tools to exploit P-States for energy management.

C-States on the contrary can lead to higher power savings as more parts of the CPU are shut down, however the transition between C-States takes significantly longer and is always directly controlled by the OS and offers no interface for the user to control it. Thus, C-States are usually deactivated for supercomputers.

## Multi-Cores and SMT

For a long time the processor frequency increased with every new generation, which was achieved by reducing the production process and adding transistors to the processor die. "Moore's law" [121]—formulated by the Intel co-founder Gordon E. Moore in 1965—states that the number of transistors of an integrated circuit doubles roughly every two years, which, for a long time, corresponded to doubling of processor performance every 18 months. However, increasing the processor frequency above a certain threshold leads to unmanageable power consumption and thermal problems. But as Moore's law still holds true, the processor vendors had room for additional transistors on the chip, which they used to add more logical units, i.e. processor cores on the die. Today essentially all server processors used in HPC systems are multi-core processors, with up to 16 cores on AMD Opterons or the Blue Gene/Q compute node.

Basmadjian and de Meer proposed a method to evaluate and model the power consumption of multi-core processors [14] with a special focus on resource-sharing and power saving mechanism.

An in-depth analysis of DVFS on Chip-Multiprocessors for several workloads ranging from web servers to scientific applications was done by Herbert and Marculescu [65]. They found that DVFS can yield in high energy savings and that, contrary to many assumptions, per-core DVFS is not necessarily beneficial as it adds more complexity to the chip. Further they showed that a hardware/software cooperation can be as efficient as a hardware only solution at a reduced hardware complexity. Lim and Freeh proposed a model to determine the minimal energy consumption possible when using DVFS [102]. Tools that exploit DVFS to reduce the energy consumption of HPC applications are presented in Section 3.4.

Many processors support simultaneous multithreading (SMT), i.e. running multiple hardware threads per core. While Intel Xeon processors support two threads per core (2-way SMT), IBM's POWER7 processor has eight cores each having four simultaneous threads. To be able to issue instructions from multiple threads in one cycle, the processors must provide extra logic like pipelines which increases the power consumption of the chip. The effect of SMT on the performance is highly application dependent as SMT may induce contention on the shared resources like memory bandwidth, caches, etc. which might actually lead to a performance degradation. Schöne et al. investigated the influence of SMT on power consumption for Intel processors [139] and found that SMT efficiency on the Nehalem and Westmere generations was not optimally implemented resulting in decreased efficiency for basically all applications. The situation highly improved on the Sandy Bridge architecture, while SMT still shows significantly higher power consumption, the performance improvements for many application outweigh the higher power consumption resulting in overall energy savings. However, on applications that cannot exploit SMT the higher power consumption yields in reduced energy-efficiency.

The server cores used in commodity clusters are general purpose processors with a lot of functional units that are not needed in the HPC context but consume a considerable amount of power. Power gating is a technique to reduce the leakage power by disabling certain parts of the processor by blocking the power supply of a logic block by inserting a sleep transistor (or gate) in series with the power supply. The gated block virtually consumes no power at all. However, switching the gate consumes a certain amount of energy and thus power gating needs to be applied carefully in order to keep the energy penalty at an acceptable degree. Lungu et al. [108] presented a power gating approach with a guard mechanism to bound the energy penalty induced by power gating. Per-core power gating in combination with DVFS was researched by Leverich et al. [98], showing that a holistic approach has a much higher saving potential than each technique applied on its own.

## Vendor Specific Developments

Most server processors are ACPI compliant, but with different implementation of the power and sleep states. Further, the processor vendors add some extra power saving functionality to their respective processors to enhance energy-efficiency, some of which we discuss now.

## Intel

Intel started to put a major focus on energy-efficiency in their Xeon server processors with the 45 nm Nehalem family [137] built with a new type of transistor, the high-K dielectric transistor with much faster switching time and reduced leakage current. The Nehalem generation includes several improvements to former Xeon processors like 5× more power states (up to 15 instead of 3) and a reduced transition time of less than 2 μs, compared to 10 μs of previous Xeon generations [58]. Additionally, new power management features like integrated power gates for the single cores were added.

The 32 nm Westmere family [93] increased the core count and cache size while staying within the same power envelope as the Nehalem family. Many architectural improvements for power efficiency were added like power gates not only for the cores (as the Nehalem provides) but also for uncore regions containing shared functional units of the processor. Overall the Nehalem and Westmere architectures showed many circuit and process innovations that led to higher performance and reduced power consumption [7].

Just recently Intel presented the Sandy Bridge architecture with new power management features [133]. It should be noted that currently no server processor with that architecture is yet released. The heart of the Sandy Bridge power management is the PCU—the Package Control Unit, which constantly collects power and thermal information, communicates with the system, and performs various power management functions and optimization algorithms. Power management agents (PMAs) connect the cores and functional units to the PCU through a power management link. These PMAs are responsible for the P-state and C-state transitions.

## AMD

AMD's latest Opteron processors [105], in HPC famous for powering the large Cray supercomputers, implement the AMD-P suite[12] of power management features which include:

- AMD PowerNow[13] with Independent Dynamic Core Technology allows processors and cores to dynamically operate at lower power and frequencies, depending on usage and workload, to help reduce the total cost of ownership and lower power consumption in the data center.
- AMD CoolCore Technology, which can reduce energy consumption by turning off unused parts of the processor.
- AMD Smart Fetch Technology, which helps reduce power consumption by allowing idle cores to enter a "halt" state, causing them to draw even less power during processing idle times, without compromising system performance.
- Independent Dynamic Core Technology, which enables variable clock frequency for each core, depending on the specific performance requirement of the applications it is supporting, helping to reduce power consumption.
- Dual Dynamic Power Management (DDPM) Technology, which provides an independent power supply to the cores and to the memory controller, allowing the cores and memory controller to operate on different voltages, depending on usage.
- Advanced Platform Management Link (APML) provides advanced controls and thermal policies to closely monitor power and cooling.
- AMD PowerCap manager gives the ability to put a cap on the P-state level of a core via the BIOS enabled, helping to deliver consistent, predictable power consumption of a system.
- Additional ACPI C-States like an enhanced halt state (C1E) and a deeper sleep state (C6).
- AMD CoolSpeed technology provides highly accurate thermal information and protection.

---

[12] http://sites.amd.com/us/business/it-solutions/power-efficiency/Pages/power-efficiency.aspx
[13] http://www.amd.com/us/products/technologies/amd-powernow-technology

## IBM

IBM's POWER7 [78] is the latest iteration of the successful IBM POWER processor family which poses the most important alternative to x86 server processors. It has been developed with a strong focus on power management [158], with a set of power management features known as EnergyScale.

EnergyScale has been introduced in the POWER6 processor [112] and further improved for the POWER7 [23]. It is mainly controlled by the Thermal and Power Management Device (TPMD), a dedicated microcontroller that was optional in the POWER6 and became an integral part of the processor in the POWER7. The TPMD collects data from various thermal and power sensors provided on the POWER7 chip and controls the power management policies of the processor. Besides the usual P–States the POWER7 introduces two new idle modes, Nap and Sleep. Nap is a mode designed for short idle periods which provides a power reduction over a software idle loop by clocking off all execution units and the L1 cache, yet the processor is still in an responsive state to resume work with low latency. The sleep state, a lower power and higher frequency standby state, is intended for longer phases of inactivity which usually do not occur in HPC.

The POWER7 offers additional adaptive energy management features [49, 50] that complement the EnergyScale firmware for more energy-efficiency. These features include per-core frequency scaling, per chip voltage settings and estimation of power consumption as well as an assistant for hardware instrumentation. Brochard et al. discussed how these features can be used to optimize the performance and energy of HPC applications on POWER7 [22].

### 2.1.2 Accelerators—GPUs

Accelerators are a hot topic in HPC for the last couple of years [47, 86], reaching a peak when the Chinese Tianhe-1A, a cluster of Intel Xeon processors and Nvidia GPUs, claimed the top spot of the November 2010 Top500 list. In the latest Top500 list of June 2012, 2 of the top 10 and 58 of the Top500, respectively, are using accelerators, most of them are powered by Nvidia GPUs. Accelerators have a significant impact in power consumption—a single high-end GPU can consume as much power as the rest of the node—and efficiency of a HPC cluster [77]. To benefit from GPUs the application must be able to utilize the particular GPU hardware—many small cores with high streaming capabilities—by a high degree of vectorization and latency hiding [111].

The influence of GPU on power consumption and efficiency of a system was quantified by Enos et al. [42]. The authors present how they integrated the measurement and visualization of GPU power consumption into a GPU accelerated cluster. They evaluate four typical HPC applications which show different speedups when using the GPUs. While the average power consumption of jobs using the GPUs were always higher, the increase in energy-efficiency highly depends on the speedup, i.e. how well the application fits on the GPUs.

DVFS is not only possible for CPUs but also for GPUs. Lee et al. [97] used DVFS and core scaling to improve the throughput of power-constrained GPUs. Their experiments show an average increase in throughput of 20% under the same power constraint.

Pool et al. developed an energy model for GPUs by measuring the energy consumed for several arithmetic and memory operations (much like Molka et al. did for CPUs [119]). With some knowledge of data flow in the hardware it is then possible to predict the energy consumption of arbitrary workloads. Such models help to investigate power consumption and energy-efficiency of GPUs even in clusters with no power measurement capabilities.

As Moore's law is still valid and more and more transistors are integrated into processors, the processor vendors started to integrate GPUs with traditional CPUs. The integrated GPUs consume much less power than their discrete counterparts as they have to fit into the processors power domain. Scogland et al. [141] studied these integrated GPUs from an energy consumption perspective. Three systems were compared for a set of applications, a multi-core system, a high-performance discrete GPU, and a low-power integrated GPU. The evaluation shows that integrated GPU is on par with the discrete GPU in terms of energy-delay product (see Section 1.3) and and delivers better performance than the multi-core CPU while consuming much less energy.

### 2.1.3 Memory

Memory is now a significant consumer of power in a HPC system and its share is likely to increase in the near future with upcoming many-core processors to keep a reasonable amount of memory-per-core and to sustain the desired memory bandwidth. However, memory power is often not addressed when discussing power saving opportunities in HPC.

There are several approaches to reduce the power consumed by the memory subsystem which we will discuss here—some on the hardware side like using energy-efficient memory modules or using the power states of the

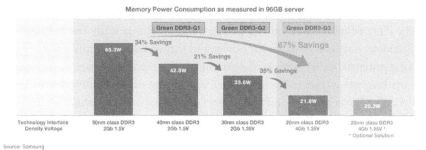

**Fig. 5.** Power saving potential in the Green DDR3 line from Samsung. Up to 65% power savings are possible using the third generation of Green DDR3 memory compared to traditional memory for a server equipped with 96 GB main memory. Source: Samsung.

memory (yes, the memory provides different power states, although they are not so easy accessible as the processor power states), other approaches try to optimize memory accesses at compile time to save energy.

Low-power memory modules—usually operating with a lower voltage of 1.35 V instead of the traditional 1.5 V— are available from several vendors like Samsung's Green Memory line[14] or Crucial's Energy-Efficient Memory.[15] Figure 5 shows the development of the Samsung Green DDR3 family by outlining the possible power savings for a server equipped with 96 GB memory.

By a continuous reduction of the production process and applied voltage power savings of up to 67% can be achieved. In the first generation the production process shrunk from 50 nm to 40 nm with constant voltage resulting in 34% power savings, in the next generation not only the production process was smaller but also the applied voltage was reduced from 1.5 V to 1.35 V which gave additional 21% savings. The latest generation also operates at 1.35 V with a 20 nm production process as 4 GB modules opposed to 2 GB modules in the previous generations. This yields in a power consumption reduction of 35% to the previous generation of Green DDR3 and an overall saving of 67%.

However, even the low-voltage memory is not constantly used by not bandwidth-limited applications (or, more precisely, phases of applications that are not memory-bandwidth limited like communication or check-pointing phases) and thus could be put to lower power states occasionally.

[14]http://www.samsung.com/global/business/semiconductor/minisite/Greenmemory/main.html

[15] http://www.crucial.com/serv er/eememory.aspx

David et al. proposed a DVFS-based memory power management in [35]. They developed a memory power model that quantifies the frequency-dependent portions of memory power and showed that a significant energy-saving potential exists when scaling memory voltage and frequency. A control algorithm was presented which observes the memory-bandwidth utilization and adjusts voltage and frequency accordingly. The evaluation on real hardware—where DVFS was simulated by altered timing settings—yielded in an average memory energy saving of 10.5% (at a maximum of 20.5%), or a total energy saving of 2.4% (5.2% max.) with a performance impact of 0.17% slowdown.

The synergistic effects of power-aware memory systems and processor DVFS was studied by Fan et al. [46]. They demonstrate that effective power-aware memory policies can enhance the impact of DVFS on processor side by lowering the memory power costs relative to the CPU and they explore how DVFS settings are influenced by various memory control policies. They further developed an energy consumption model derived from hardware counter information and show that the memory access pattern of the application is highly relevant for an optimal DVFS setting.

However, often the memory idle periods are rather short and thus lower power states cannot be used efficiently as the transition times between memory power states are too long. Huang et al. [76] proposed a method to efficiently exploit memory power states by reshaping the memory traffic, consolidating memory accesses and combining multiple short idle states into fewer longer ones. Up to 38% additional energy savings are possible with this approach combined with traditional power-management techniques.

The group around Mahmut Kandemir and Victor Delaluz at Pennsylvania State University focused on compiler-level optimizations to reduce the energy consumed by the memory subsystem by an optimized data layout and order of operations. In [36] they investigate the optimal data placement on a multi-bank memory system. The approach is to cluster arrays with so-called similar lifetime patterns in the same memory banks and use lower power states for these banks when the lifetime of those arrays ends. A reordering of loop executions to keep memory banks longer idle is presented in [81]. The loops are classified using an access pattern classification and with loop iterations are reordered according to that classification. This transformation works for both parallel loops and loops that run sequentially due to data dependencies. Overall about 33% of the memory subsystem energy can be saved. They further studied the influence of loop optimizations on the energy consumption memory systems in [80]. The influence of several loop transformation techniques—loop fission and fusion, linear loop transformations, and loop tiling—was investigated for array-dominated codes and yielded in significant energy savings with competitive run-time.

Kim et al. [83] studied the effects of high-level compiler optimizations on the energy consumption of the memory subsystem for both instruction and data accesses. While most optimizations improve data locality and reduce the energy needed for data accesses they have a negative influence on instruction access yielding in a higher energy consumption there. So techniques that consider both data and instruction access have to be researched and developed.

### 2.1.4 Network and Storage

Network and storage never played a critical role for power optimization in the HPC context. However, both become more important as the system sizes increase. While the power consumption of a single node stays constant or even shrinks, the network and the I/O-subsystem have to scale with the machine.

Figure 6 shows the distribution of HPC interconnect families (Fig. 6a) and and their corresponding share in performance (Fig. 6b) of the Top500 list of November 2011. The dominant interconnect families in HPC with a total share of over 85% are Ethernet (44.8%) and InfiniBand (41.8%). Performance wise the custom interconnects are leading with a performance share of 24.1% at a systems share of 5.8%. InfiniBand's system and performance share is nearly the same whereas Ethernet's performance share is only half of its system share.

Ethernet adapters support different link speeds which yield in power savings of $O(1\,W)$ for Gigabit Ethernet to $O(10\,W)$ for 10 GbE adapters (depending on the model). That does not sound much, but on a cluster with several thousand nodes that is a substantial saving. InfiniBand, the second big player in HPC interconnects provides the possibility to shut down unused lanes in order to reduce the power consumption.

Torsten Hoefler from the National Center for Supercomputing Applications (NCSA) at the University of Illinois is one of the leading experts in the area of energy-efficient HPC interconnects, giving a good introduction into the field in [69], covering both software and hardware aspects of energy-efficient networking. Hoefler et al. investigated the energy—performance tradeoff in topology-aware mapping of MPI tasks to a HPC system in [70]. An optimal mapping can be searched for two objective functions, the maximum congestion and average dilation. While congestion is a measure for

## Interconnect Family System Share

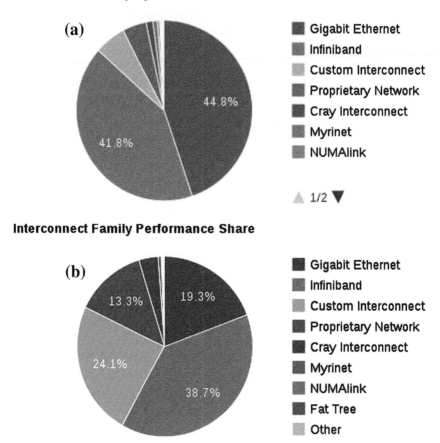

**Fig. 6.** Overview of the share and the corresponding performance of different types of interconnects for the June 2012 Top500 list. Source: Top500.org.

performance, the dilation, i.e. the number of "hops" of a message from the sender to the receiver, is an indicator of the energy-consumption of the network. The regarded mappings showed different results for the two metrics, some are better in reducing the congestion while others minimize the dilation, so it is up to the developer whether he/she cares for performance or energy consumption.

Lui et al. investigated the influence of remote direct memory access (RDMA) on the power consumption of RDMA-capable networks like InfiniBand compared to traditional communication over TCP/IP [104]. While the power consumption of RDMA during communication is significantly higher it still yields in an increased energy-efficiency especially in communication intensive phases. This is due to a reduced interaction of network interface and CPU and memory subsystem, resulting in fewer CPU cycles needed for RDMA communication.

The situation is even worse for the I/O-subsystem. While there is some research on how to reduce the power consumption of hard disks, compute nodes in HPC systems usually are not equipped with disks as the I/O system is provided by large-scale parallel file systems running on their own file servers. There has not been a lot of research in that field yet.

The easiest way to reduce the power consumption of the storage subsystem is replacing traditional mechanical hard disks by solid state drives (SSDs), which are advantageous in basically every way— providing a high number of IOPS (Input/Output Operations Per Second) at very low latency and little power consumption. The points where SSDs fall short are capacity, which is not as important in HPC as bandwidth and latency requirements necessitate a large number of disks, and, more importantly, price as SSDs still are far too expensive for production-level HPC storage.

Kunkel et al. [92] proposed a model to calculate the energy demand of scientific data of its complete life cycle, from creation over storage on a fast file system till archived or disposed. They further present several ideas how file systems can be extended to provide the required information.

## 2.2 Outlook

A lot of research is done to design the components for the future. However, some promising ideas from nano-electronics will take at least 10 more years before they will result in production. Nevertheless, there are certain developments that can result in the next generation of energy-efficient processors, memory, and network interfaces [156].

### 2.2.1 CPUs

Intel is currently producing processors based on 22 nm technology and develops fabs for a 14 nm production process which should start in 2013. In addition, Intel has already started researching for even smaller production processes of 10–5 nm. Such small lithography will require a lot of new technology to keep the leakage current at a manageable level, e.g. tri-gate transistors [82]. Tri-gate transistors are 3D components with a larger area for electrons to travel and thus reducing the leakage power. Compared to traditional transistors the tri-gate transistors are up to 37% faster and using about

50% less power. The Ivy Bridge processors built with a 22 nm production process are first processors to include tri-gate transistors.

Apart from using increasingly smaller lithography there are new concepts of processing that could be potential game-changers for energy-efficient architectures. The most promising developments are resilient computing and near-threshold computing, which allow the operation of the processor at a very low voltage level that is not feasible for current microprocessors.

At the first European "Research @ Intel" day 2010, Intel presented a prototype of a RISC processor, called "Palisades," that enables the so-called resilient computing, i.e. the processor operates at a very high voltage (for performance) or very low voltage (for energy-efficiency). At both operating modes the operations of the processor likely produce a wrong result, thus current processors operate within strict voltage limits. The Palisades processor can operate at extreme voltage levels as it has some electrical error detection and correction implemented, which ensures that only correct values are written to cache or memory. The Intel team run several thousand of tests and compared with a Palisades version that runs error free at a certain power level, the resilient version is up to 41% faster at the same power level or reduces the power consumption by 21% delivering the same performance. However, this technique is far from being used in a commodity processor.

The second hot topic in processing is near-threshold computing (NTC) as proposed by Dreslinski et al. [40]. This means that the system operates at a voltage that is nearly the threshold voltage of the transistors, i.e. the voltage at which the transistor is turned on. Usually much higher voltages are supplied to minimize errors in unintended state switches of the transistor. The authors claim that NTC could reduce the energy consumption of future computing systems by a factor of 10–100. Intel presented a first NTC prototype processor, code-named "Claremont," at the Intel Developers Forum 2011. While this technological study is not planned to become a product it might lead the way for future microprocessors.

### 2.2.2 Memory

A great amount of research is done in the field of memory leading to several approaches competing to become the next generation of HPC memory replacing the current DRAM. Phase-change memory[16] (PCM) is a non-volatile random-access memory based on the state-switching (crystalline and amorphous) of certain alloys, mostly chalcogenides. Wong et al. give

---

[16] http://www.micron.com/abo ut/innovations/pcm

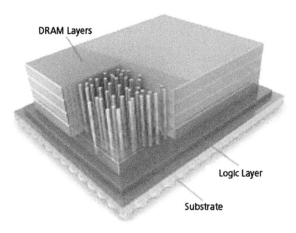

**Fig. 7.** A scheme of the HMC design. Source: Micron.

an overview of phase-change memory in [163]. They present various materials that could possibly be used to produce PCM devices and discuss their electrical and thermal properties and the corresponding impact on the design of a PMC device. A possible solution is a PCM main memory with embedded Modules as cache. Such a solution have been proposed by Qureshi et al. [131] and the energy-efficiency of a hybrid PCM/embedded DRAM system was analyzed by Bheda et al. [19]. Those experiments show that a 3x speedup and a 35% improvement in energy-efficiency is possible.

In 2011, Micron presented the Hybrid Memory Cube[17] (HMC) [129], a novel 3D memory architecture. It consists of several layers of DRAM stacked on a logic layer with several buses to the memory layers as depicted in Fig. 7.

The HMC is expected to provide 15x increased bandwidth while saving up to 70% energy per bit compared to current DDR3 memory. To force the development and applicability of the HMC, several companies formed the Hybrid Memory Cube Consortium[18] with the aim to define "an adoptable industry-wide interface that enables developers, manufacturers and enablers to leverage this revolutionary technology." At the end of 2011, Micron and IBM announced that they will build products based on the HMC technology.[19]

---

[17] http://www.micron.com/about/ innovations/hmc
[18] http://hybridmemorycube.org/
[19] http://www-03.ibm.com /press/us/en/pressrelease/36125.wss

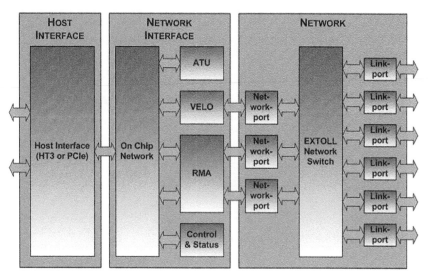

**Fig. 8.** The EXTOLL network architecture. Source: EXTOLL.

Other approaches to redefine memory are the Memristor, IBM's Racetrack memory, and floating-gate devices. The Memristor—the memory resistor—was first described in 1971 by Chua [27] as the fourth fundamental circuit element beside the resistor, the capacitor, and the inductor. In 2008, a team from the HP Labs was able to show that memristance is actually possible in nanoscale systems [151]. Several teams are currently working on products based on memristors which need a much smaller area than traditional transistors and thus can be applied in a wide range of products including non-volatile memory with a high density. Although at an early experimental stage, IBM's Racetrack memory [128] and novel floating-gate devices which could be used to unify volatile and non-volatile memory devices [138]. Both technologies however might play an important role in future memory architectures if they will evolve into productions.

### 2.2.3 Interconnects

Interconnects gain an increasing interest in academic and industrial research. A new network architecture for HPC has been presented by EXTOLL[20], its highly modular architecture is shown in Fig. 8.

---

[20] www.extoll.de

The EXTOLL adapter combines the host interface, the network interface, and the networking functionality on one chip, making it highly power efficient. It integrates a separate switch to handle different message types with very specialized hardware, e.g. a special low-latency communication engine called VELO (Virtualized Engine for Low Overhead) [103] is integrated which handles small messages much more efficiently than typical DMA (direct memory access) approaches. It further provides dedicated hardware for remote memory accesses (RMA) and an address translation unit (ATU). The whole on-chip network is attached to the host processor by either PCI Express or AMD's Hyper Transport, allowing a very fast connection to the processor. Six network ports are provided by the EXTOLL adapter forming a 3D torus network. The original adapter was FPGA (Field-Programmable Gate Array—an integrated circuit that can be configured by the user) based [127, 54] but is going to be replaced by an ASIC (an Application-Specific Integrated Circuit—an integrated circuit for special purpose usage), making it even more efficient in terms of power consumption and costs.

Next-generation HPC interconnects will most likely be built upon photonics, i.e. information will be processed as optical signals instead of electrical signals. Optical connections have several advantages over electrical transmissions. Optical networks are basically unrestricted of the diameter of the network, which can be spanning several hundred meters for large-scale systems. They further provide a very high bandwidth and a low latency, making them the ideal candidates for HPC interconnects. Additionally, optical components typically consume considerably less power then their electric counterparts. There are still some problems to solve—mainly technological issues in the optical components—before real optical networks conquer the space of HPC. Several research projects investigated certain kinds of application for photonics.

Kodi and Louri researched reconfigurable photonic networks for HPC [90, 91]. Their proposed solutions increase the performance of the network by dynamic bandwidth relocation based on the communication traffic pattern and simultaneously decrease the power consumption by using dynamic power management techniques. Researchers at IBM and Corning developed the Osmosis project, the Optical Shared Memory Supercomputer Interconnect System [118], to study whether optical switching technologies can be used in HPC systems.

Not only inter-node networks are of interest for the photonics community, but also intra-node networks, so-called NoCs (Network-on-Chips), i.e. the data-paths between the different cores of a processor and between cores and memory have been investigated for several use-cases. Kim et al. [84] gave an overview of the different technologies that are available and investigated the impact of the various technologies on the system design for on-chip communication. A new monolithic silicon-photonic technology based on standard CMOS processes was introduced by Batten et al. [15]. They explored the feasibility of this cost and energy-efficient technology for processor-to-memory networks. Different multiplexing strategies for photonic NoCs are presented by Fey et al. [48] and they discuss the benefits and drawbacks of using the different technologies in HPC architectures.

## 2.3 Energy-Aware Architectures
### 2.3.1 Trends in x86 Clusters

A hot topic for the reduction of the power consumption of x86 clusters is the replacement of power-hungry air cooling by hot-water cooling with heating reuse, i.e. 60 °C (140 °F) warm water is used to transport the heat away from the processors. In the end the water has a temperature of about 85 °C (185 °F), which is warm enough to efficiently use the water for other purposes, e.g. heating the offices in the building. With traditional water cooling methods, where significantly colder water is used, the water at the end of the cooling circuit is still too cold to be used for other purposes and has to be cooled down again for the next cycle. Water has a 4000 times increased heat capacity compared to air, which makes it an ideal medium for cooling down supercomputers. The AQUASAR system [166], a collaboration of IBM research Zurich and the ETH Zurich, showed a first hot water cooled supercomputer prototype with heat reuse for building heating.

Build upon the concepts developed in the AQUASAR project, Super-MUC[21] [10], a 3 PFlops IBM supercomputer at LRZ Munich, proves that hot water cooling is possible for a large-scale system. The system with 155,656 Intel Xeon processor cores in 9400 compute nodes entered the June 2012 Top500 list as rank #4.

The most power hungry components like processors and memory are directly cooled by water with an inlet temperature of up to 40 °C. The hot water at the outlet (approximately 75 °C) will be used to heat the adjacent office building and is then cooled down "for free," i.e. no cooling equipment except cooling towers are necessary, by the ambient air for the next cycle in the cooling process. Further, the operation of the system is closely monitored

[21] http://www.lrz.de/services /compute/supermuc/

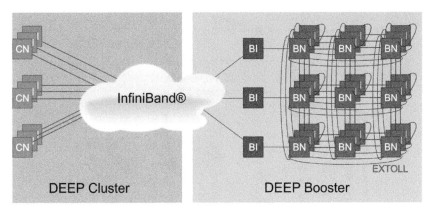

**Fig. 9.** The hardware architecture of the DEEP project.

and supported by an energy-aware software stack from IBM (see Section 3.4), enhancing the efficiency even more.

The DEEP project,[22] funded by the European Union (EU) in the Seventh Framework Programme (FP7) as one of three Exascale computing projects, proposes and evaluates a novel HPC architecture, the so-called cluster-booster architecture [41] shown in Fig. 9.

A traditional cluster with Intel Sandy Bridge Xeon processors is connected to what is called a booster (or cluster of accelerators), equipped with Intel "Knights Corner" (KNC) processors of the MIC (Many Integrated Cores) platform which are connected by a 3D Torus interconnect. The cluster nodes (CN) can communicate to the booster nodes (BN) via booster interfaces (BI) that are connected to the cluster with InfiniBand and to the booster nodes with the EXTOLL network.

This approach combines powerful nodes with high single thread performance for low to medium scalable kernels and massively parallel nodes for highly scalable kernels. Applications will be enabled to offload their highly scalable kernels to the booster nodes while the cluster nodes can concurrently work on less scalable parts. This requires new programming paradigms that will be developed in the project.

The system is designed with a strong focus on energy-efficiency, being based on the hot liquid cooled Eurotech Aurora[23] solution and integrating the efficient EXTOLL network (see Section 2.2) to interconnect the booster nodes. The cooling concept is outlined in Fig. 10. Further, energy-aware

---

[22] www.deep-project.eu

[23] http://www.eurotech.com/en/hpc/hpc+solutions/data+center+hpc/Aurora+Systems

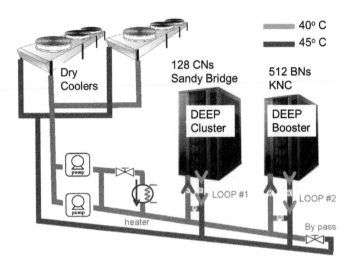

**Fig. 10.** The cooling concept of the DEEP project.

software solutions are planned to reduce the power consumption of cluster and booster nodes when not fully utilized.

### 2.3.2 SoC-Based Systems

For many years supercomputers used the most powerful processors available to deliver maximal performance at the expense of increased power consumption. The number 1 system in 2002, the Japanese Earth Simulator based on NEC vector processors, consumed an incredible 18 mW to deliver 35.6 Teraflops.

Wu-Chun Feng, the creator of the Green500 list, initiated the Supercomputing in Small Spaces (SSS)[24] project at Los Alamos National Laboratory. The aim was to develop a power and space efficient cluster computer built on commodity hardware from the embedded systems sector. This resulted in the Green Destiny cluster [159], a 240 node cluster with Transmeta TM5800 processors that delivered 100 Gigaflops while consuming only 3.2 $\tilde{k}W$ of power on a floor space of 1 m$^2$.

This project started the investigation of the usability of embedded Systems-on-a-Chip (SoCs) [130], i.e. chips with integrated processor, memory, and network, in HPC systems. IBM brought up the most prominent

---

[24] http://sss.lanl.gov/

and successful implementation of this approach, the Blue Gene family of supercomputers.

### 2.3.3 IBM Blue Gene Family

The Blue Gene approach followed the famous Einstein quote "Everything should be made as simple as possible, but no simpler." So no bleeding edge technology was used, instead the engineers used a standard and reliable PowerPC SoC from the embedded systems area. Thus, the focus could be put on low-power consumption and efficient communication networks, incorporating a 3D torus network for point-to-point communication and a tree network for collectives. A dense packaging could be achieved with 1024 nodes per rack, divided in two midplanes with 16 node cards. Each node card includes 32 compute cards (with one chip each) and an integrated network to connect the compute cards. This packaging, shown in Fig. 11, was maintained through all generations.

To avoid OS jitter as much as possible, the nodes are running a very small compute node kernel instead of a full Linux distribution.

The Blue Gene family held several top positions in both the Top500 and the Green500 lists. As of June 2012, both lists are dominated by Blue

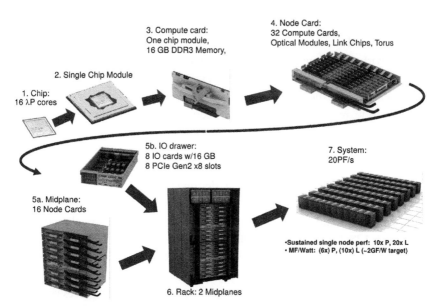

**Fig. 11.** The Blue Gene/Q setup for the Sequoia installation. Ninety six racks with 1.6 million cores deliver 16 PFlops within an 8 MW power envelope. Source: IBM.

**Table 1** Blue Gene Family Comparison. The Peak Performance and Total Power are given for the largest installation.

| Property | BG/L | BG/P | BG/Q |
|---|---|---|---|
| Processors/chip | 2 PowerPC 440 | 4 PowerPC 450 | 16 A2 |
| Processor frequency | 0.7 GHz | 0.85 GHz | 1.6 GHz |
| Processor Voltage | 1.5 V | 1.2 V | 0.7 V |
| Coherency | Software | SMT | SMT |
| L2/L3 Cache size (shared) | 4 MB | 8 MB | 32 MB |
| Main memory/processor | 512 MB/ 1 GB | 2 GB/ 4 GB | 16 GB |
| Memory Bandwidth | 5.6 GB/s | 13.6 GB/s | 42.6 GB/s |
| Peak Performance/node | 5.6 GF | 13.6 GF | 204.8 GF |
| Torus Network Bandwidth | 2.1 GB/s | 5.2 GB/s | 40 GB/s |
| Torus Network Latency: HW | 0.5 $\mu$s | 0.5 $\mu$s | 0.3 $\mu$s |
| Torus Network Latency: MPI | 4 $\mu$s | 3 $\mu$s | 2.5 $\mu$s |
| Peak Performance | 509 TF (104 k) | 1 PF (72 k) | 20 PF (96 k) |
| Total Power | 2.3 MW (104 k) | 2.3 MW (72 k) | 8 MW (96 k) |

Gene/Q systems, led by Sequoia, the 96-rack system at Lawrence Livermore National Lab (LLNL).

The system details of the three generations of Blue Gene are given in Table 1. The Blue Gene/P was just a minor technical update of the Blue Gene/L, with an update of the CPU architecture from PPC440 to PPC450, but the network topology and the systems organization stayed the same. Significant performance improvements were reached through higher density (4 cores/node instead of 2) and architectural improvements. Jugene, the 72-rack Blue Gene/P installation at JSC, pictured in Fig. 12, was the first European petascale supercomputer.

Unlike the previous generation, the Blue Gene/Q includes a lot of cutting edge technology. The completely new designed compute chip [63] is a 17-core chip, 16 cores are usable for computation and the 17th core runs the operating system and service tasks. All cores offer 4 hardware threads per core and a 128-bit vector unit called QPX. Novel features like hardware transactional memory and thread level speculation offer new possibilities for parallelism and hence an increase in efficiency. A new 5D torus interconnect [26] for both point-to-point and collective communication replaces the two distinct networks. The Blue Gene/Q is entirely water cooled, allowing a very dense packaging of the powerful nodes.

**Fig. 12.** Jugene, the 72-rack Blue Gene/P installation at Juelich supercomputing centre. With nearly 300,000 cores, this system was the most parallel computer in the world from 2009 to 2011.

A novel resource management has been established [24] to maximize the utilization of the machine. Blue Gene/Q partitions for a job are allocated as blocks which consist of a certain number of nodes. The minimum number of nodes depends on the number of IO drawers available for a midplane and can be quite high. The new resource manager allows so–called sub-block jobs, i.e. multiple jobs can run in one block. This minimizes the number of idle nodes and thus improves the energy-efficiency a lot.

Besides the DEEP project the European Union funded a second Exascale project in the 7th Framework Programme which evaluates the suitability of low-power embedded technology for an Exascale machine. The Mont-Blanc project[25] tries to overcome the obstacles in hardware like maintaining a reasonable memory bandwidth and software, e.g. providing a programming environment, in using ARM processors, probably with an integrated GPU like the Nvidia Tegra[26] in a highly scalable supercomputer.

A similar approach is taken by HP with the Project Moonshot [68] and the HP Redstone server development platform. The 4U chassis can be packed 288 quad-core ARM chips or low-power x86 Intel Atom processors

[25] http://www.montblanc-project.eu/
[26] http://www.nvidia.com/object/tegra.html

and offers integrated shared power distribution and cooling as well as a high-performance interconnect. While HPC is not the primary target of that platform it might evolve into a new HPC paradigm if Project Moonshot turns out to be scalable for a range of different workloads.

### 2.3.4 Application-Centric Architectures

Special machines have also been built which are optimized for a certain type of application, the so-called hardware-software co-design [75]. Those systems excel in their respective domain at the cost of general purpose usability. However, some systems provide the necessary tools and libraries— especially a MPI library, to maintain the ability to execute non-specialized code. Co-designed supercomputers often contain special parts and custom interconnects, making them relatively expensive in design and development.

Probably the most prominent example for hardware-software co-design in the HPC space is QPACE[27] (Quantum Chromodynamics Parallel Computing on the Cell) [61], a massively parallel and scalable computer architecture optimized for lattice quantum chromodynamics (LQCD).

The QPACE project was developed in a collaboration of several universities under the lead of the University of Regensburg, the Juelich Supercomputing Centre (JSC) and the IBM research & development lab in Böblingen, Germany. This system was developed specifically for efficient solution of the QCD problem, which is mainly solving huge sparse linear equations. The heart of QPACE is the very energy-efficient PowerXCell 8i processor, a variant of the Cell BE developed originally for the PlayStation3 by Sony, Toshiba, and IBM, with support for more memory and improved double precision performance. The PowerXCell is a 9 core processor with one Power Processing Element (PPE) and 8 Synergistic Processing Elements (SPE) which are very efficient for vector processing. The nodes are connected by a custom FPGA-based 3D torus network which is ideal for QCD.

Power efficiency was a top design constraint and several strategies have been applied to reduce the systems power consumption [11]:

- The most power-efficient components have been chosen and the number of power-consuming components has been reduced as much as possible.
- Voltages have been reduced as much as possible.
- The system is cooled by a custom water cooling system.

[27]http://www.fz-juelich.de/portal/EN/Research/InformationTechnology/Supercomputer/QPACE.html

**Fig. 13.** The 4-rack QPACE installation at JSC.

One power optimization feature is an automatic node-level processor voltage tuning process where every processor runs a synthetic benchmark at every voltage level and stores at the lowest voltage level for which the processor still operates correctly.

There are two identical QPACE installations, one at JSC and one at the University of Wuppertal. Both installations deployed 4 racks with 32 backplates, each with 32 nodecards, Fig. 13 shows the QPACE installation at JSC. With the total 1024 PowerXCell processors a peak performance of 100 Teraflops can be reached with an aggregated power consumption of 140 kW. The resulting power efficiency was 723 Megaflops/W which led to the top sport of the November 2009 Green500 list (and rank 110 in the Top500 list), which it defended also on the June 2010 list.

Green Flash,[28] a supercomputer for climate predictions [160], developed at the Lawrence Berkeley National Laboratory, builds upon the tiny Tensilica XTensa LX, an extensible embedded processor from the mobile industry which offers a very high floating-point efficiency [39] instead of using

---

[28] http://www.lbl.gov/CS/html/greenf lash.html

high-performance processors like the PowerXCell. Although optimized for climate predictions, Green Flash will offer standard tools and libraries enabling it to run most large-scale applications.

Anton [145], a supercomputer specialized for molecular dynamics (MD), follows yet another approach by building upon custom designed ASICs. The ASICs contain two main functional units, the main MD calculations of electrostatic and van der Waals forces is performed by the high-throughput interaction subsystem (HTIS) [96]. All other calculations like bond forces and FFT are performed on a flexible subsystem [94], which contains four Tensilica cores (similar to the ones used in Green Flash) and eight geometry cores, which are specialized but programmable SIMD cores and thus suited for vector operations. The ASICs are connected by a 3D torus network, a topology that has proven its suitability for HPC in many machines now.

## 3. SOFTWARE ASPECTS OF ENERGY-AWARE HPC

This section focuses on the software aspects of energy-aware HPC. Software has a high effect on power consumption and energy-efficiency of a supercomputer on many scales [148]. From the cluster management software over the application(s) running on the system down to the system software running on the node.

Several tools have been developed that exploit hardware power management features in order to exploit potential to reduce the energy consumption of applications. Many HPC centers nowadays put a focus on power consumption when procuring a new system without considering the efficiency of the applications running on their machines. HPC centers could develop an energy-aware accounting instead of the pure CPU-cycle based accounting that is currently the standard. This is hard as it lacks suitable energy measurement devices. However, new systems are becoming equipped with more and more sensors which could allow an implementation of an energy-aware accounting. This certainly would motivate the application developers to increase the energy-efficiency of their applications and the research on more energy-efficient algorithms.

### 3.1 Vendor-Specific Tools

Many system vendors provide their own powerful tools to measure and control the power consumption of their respective systems—however, those tools are often tailored for enterprise data center and not really suited for

HPC data centers as the measurement intervals are often in the range of seconds to minutes range which is too coarse for HPC requirements.

IBM offers the Active Energy Manager[29] (AEM) [2] as part of the Systems Director suite for their Power systems. AEM offers monitoring and control of power and thermal information proving a single view across multiple platforms, enabling a more complete view of energy consumption within the data center. AEM is not restricted to IBM servers but supports a wide variety of power distribution units (PDUs), meters, and sensors via integration with a wide variety of infrastructure vendors. It includes management function like power capping and setting of power saving modes on a single machine or groups of machines.

The dedicated IBM HPC software stack with the xCAT cluster management software and the LoadLeveler scheduling system puts a focus on energy-efficiency. Ad hoc power management will be possible with xCAT, e.g. when other systems need temporarily more power by setting a power capping value and a power saving mode. It further displays information like power consumption and temperature values, CPU usage, and fan speed.

LoadLeveler will get several new energy management features. It sets idle nodes with no scheduled workload in the lowest power state to be able to immediately use it when a fitting job is submitted. When a job runs a tag is generated for the job and an energy report is produced which contains the elapsed time and the amount of energy consumed by the job. Additionally, the possible energy savings if running at another frequency are reported. The user can then resubmit the job with the corresponding tag to run at the optimal frequency.

Intel provides tools for both the data center level to manage a whole bunch of servers as well as tools to work at system level [132]. Intel's Data Center Manager[30] is pretty much similar to IBM's Active Energy Manager and can be used for monitoring, trending, and control of power consumption and environmental aspects at scales from a single rack up to the whole data center without the necessity to install software agents on the systems. It supports real-time power and temperature monitoring and issues alarms if the values exceed custom thresholds. Historical information are stored for one year to enable the user to examine the effect of improvements. A vast set of control mechanism is provided to apply power consumption policies

---

[29] http://www-03.ibm .com/systems/software/director/aem/
[30] http://software. intel.com/sites/datacentermanager/index.php

and power capping by time (e.g. at night when power is cheaper the power consumption limit might be higher) and workload.

HP's Power advisor [66] helps in the planning phase of a new system by assisting in the estimation of power consumption and proper selection of components including power supplies at a system, rack, and multi-rack level.

## 3.2 Scheduling

On large-scale HPC systems in general multiple applications are running concurrently, usually on distinct parts of the machine, although, especially on hybrid CPU/GPU clusters, some jobs might share a node. Rarely one application harnesses the whole machine.

Users submit a job script, containing the requested resources, the application to run, and its parameters, to a batch system. Batch systems usually provide two tools that take care of running all jobs in an collision-free order, the scheduler and the resource manager which work closely together.

Batch systems usually provide text-based outputs to examine the state of the queues and the currently running jobs with their properties like wall-clock time limits and allocated resources. Such command line tools however are not feasible to monitor the system and batch load for large-scale machines with several thousands of processors. LLView [53], developed at JSC, is a client-server based tool to monitor the utilization of batch system controlled clusters, supporting a wide range of batch systems. It is installed on many supercomputers around the world.

Figure 14a shows a screen-shot of LLView for Jugene, the 72-rack IBM Blue Gene/P with 294912 processors at JSC. On the left side the 9 rows with 8 racks each are displayed and on the right side on top a table view of the currently running jobs, in the middle the status of the queue and the utilization of the system and at the bottom a prediction of the scheduler. Figure 14b shows a detailed view of four racks, each of the small rectangles stands for one Blue Gene/P node card with 32 compute cards each with 4 cores. Figure 14c details the utilization view, which shows the utilization of the machine as well as the total power consumption for the last 3 days. Further the current power consumption is displayed.

Unfortunately, scheduling is a hard problem and can only be solved heuristically. This has motivated a wide field of research for efficient scheduling strategies in order to maximize the utilization of the machine.

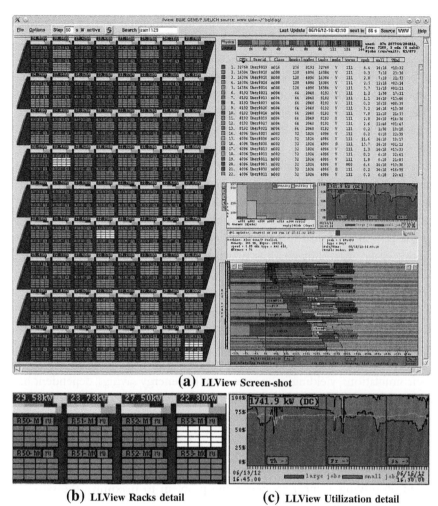

(a) LLView Screen-shot

(b) LLView Racks detail  (c) LLView Utilization detail

**Fig. 14.** Screenshot of LLView, the system monitoring software from JSC (a), showing Jugene, the 72-rack Blue Gene/P at JSC. Highlighted are the power consumption display per rack (b) and the overall utilization and power consumption (c).

Mämmelä et al. presented an energy–aware job scheduler for HPC data centers [125] within the Fit4Green project.[31] They investigated several scheduling strategies, including the simple FIFO (First In, First Out) and the first-fit and best-fit backfilling strategies and compared these with energy-

---

[31] www.fit4green.eu

aware implementations. Simulation results show that the best-fit backfilling algorithm yields the best results and energy savings of 6–8% are possible with the energy-aware strategy. They further implemented the energy-aware best-fit backfilling algorithm and tested it on a cluster at the Juelich Supercomputing Centre with computational (Linpack), memory (STREAM), and IO intensive (Bonnie++) benchmarks. The comparison with the default Torque scheduler showed energy savings of 6% at only a slightly increased overall time (+0.6%).

A group at the Barcelona Supercomputing Center investigated the scheduling policies for power-constrained HPC centers [43]. Etinski et al. proposed a power budged guided scheduling policy which uses DVFS to reduce the power consumption per job to increase the overall utilization of the machine. Simulations with four workload traces—to simulate real-world conditions—showed up to two-times better performance at a reduced CPU energy of up to 23% compared to power budget scheduling without DVFS.

The same group explored a utilization driven power-aware parallel job scheduling [45]. Here DVFS is only applied if the utilization of the system is below a certain threshold to keep the performance impact on the jobs as low as possible. Again they used workload traces from production systems for their simulations, yielding in an average of 8% energy savings, dependent on the allowed performance penalty. They further did simulations with the same workload on a larger system, showing that both performance and energy-efficiency increase significantly.

Power-aware scheduling policies for heterogenous computer clusters have been researched by Al-Daoud et al. [3]. They propose a scheduling based on solving a linear programming problem to maximize the systems capacity. The policy takes the power consumption of each machine into account to find the most efficient machine for each job, resulting in maximum energy savings.

## 3.3 Power Measurement and Modeling

### 3.3.1 Power Measurement

Power measurement of supercomputers is a difficult, especially for fine-grained measurements at component level. Usually only small test systems have been equipped with power meters, the big HPC systems contained none or only very limited power measurement capabilities. However, as popularity of power management increases, it is expected that more HPC systems with integrated power measurement capabilities will be developed.

A widely deployed framework for power measurements and profiling is PowerPack [56]. PowerPack is able to isolate the power consumption of single components like processors, memory, network, and disks and correlate the values to application functions at a very fine-grained level. The framework can be applied on clusters with multi-core multiprocessor nodes as they are common in HPC. So very precise energy profiles of scientific applications can be generated and the influence of DVFS can be investigated at an unprecedented level of detail.

Molka et al. measured the costs for operations and data movement by measuring the energy consumption of micro-benchmarks with a defined load, calculating the costs per operation [119]. This method can be applied for all kinds of processors and the resulting information can be correlated to hardware counter measurements to determine the energy consumption profile of an application.

The Blue Gene/P is one of the few large-scale machines with integrated power measurement capabilities. Hennecke et al. [64] described the power measuring capabilities of the Blue Gene/P at various levels in the system from a node card to a rack. The power measurements on the rack level have been integrated in the LLView monitoring tool (see Section 3.2). Figure 14b shows how the power consumption per rack is displayed by LLView and Fig. 14c shows the power consumption correlated to the utilization of the whole machine. In the eeClust project-power measurement and profiling methods for the Vampir performance analysis tool were developed. This project is presented in detail in Section 4. Additionally power measurements are often possible with the vendor specific tools presented in Section 3.1.

### 3.3.2 Power Consumption Modeling

Since measuring power data at scale is extremely difficult or even impossible, several groups are working on estimating the power consumption of HPC systems. Most are looking at hardware performance counters and correlate the data to measured power values to develop a model for the whole system or specific components. For hardware counter measurements, usually libraries like PAPI [154], which supports CPU counters as well as other sources like network units or all kinds of sensors through components, are used. PAPI offers both a high-level interface and a low-level interface. The low-level interface offers direct access to the hardware counters. However, using the low-level interface is not portable as the available counters vary from platform to platform. For the high-level interface PAPI predefines some counters that are available on most platforms (like PAPI_TOT_INS for the total instructions) and allows a portable access to such counters. Here the underlying low-level counters may vary from platform to platform but that is transparently handled by the PAPI library.

Lively et al. [106] developed application-centric predictive models of hybrid (MPI/OpenMP) scientific applications on multi-core systems to predict the run-time and the power consumption of the system, CPU, and memory. The model is based on a number of hardware counters—cache accesses, branch instruction, translation lookaside buffer (TLB) misses— that are recorded, but it only considers the relevant counters—it ditches counters below a certain threshold as they have no significant influence in the program behavior—and performs a multivariate linear regression over the remaining counters to predict the power consumption of the system as well as on component level within a 3% error range. Portability to other platforms is provided as the model is based on PAPI high-level counters.

A component-based infrastructure for automatic performance and power measurement and analysis based on the TAU performance analysis toolkit is presented by Bui et al. [25]. They too modeled the performance and power consumption with hardware counter information and extended the TAU PerfExplorer analysis tool to use these models for analysis.

A collaboration of Chalmers University in Sweden and Cornell University used the hardware counter measurements of micro-benchmarks to create application independent power models and thus can be used for future applications as well. The results are used to develop a power-aware thread-scheduler [146] and an intelligent resource manager [60].

Power consumption modeling and estimation is not limited to CPU and memory but can also be applied for GPUs. Nagasaka et al. [126] modeled GPU power consumption with hardware counters. Linear regression models are used in this case to model the power consumption of 49 CUDA kernels with an error rate of about 5% for most kernels. The authors found that for some kernels which are making heavy use of the texture unit the predictions are far too low as suitable hardware counters for that unit are missing.

An integrated GPU power and performance model (IPP) to calculate the optimal number of active processors has been presented by Hong and Kim [71]. They argue that if the memory bandwidth is fully utilized, adding more cores will not increase the efficiency of the application. IPP does not rely on performance counters, it uses an analytical model to predict the optimal number of cores that results in the highest performance per Watt. Evaluation shows that energy savings of up to 20% are possible for bandwidth-constrained kernels.

## 3.4 Tools to Optimize the Energy Consumption of HPC Applications

Most HPC applications use MPI for communication. Due to workload imbalances the processes usually do not reach the synchronization points at the same time, which yields in idle periods, or slack, on the processes that have to wait. A big potential to reduce the energy consumption of an application is to use DVFS to either save energy in those phases or reduce the frequency of a process so slack does not occur at all.

The group of Freeh and Lowenthal did a great amount of work in the area of energy-efficient MPI programs. They explored the energy-time tradeoff in MPI Programs on a power-scalable cluster [52] by an analysis of the NAS benchmarks. The influence of memory and communication bottlenecks on the power consumption was evaluated and exploited by using DVFS. An interesting observation is that for some programs it was possible to reduce the time and the energy consumption by using more processes at a lower power state. They also present several approaches to save energy in MPI programs. Application traces are used to divide the program into phases and for each phase the optimal frequency is chosen by a heuristic based series of experiments [51]. Using multiple energy gears, i.e. a set of frequencies, yields to better results than any fixed-frequency solution. Another approach uses a linear programming solver which calculates an optimal schedule from the applications communication trace and the clusters power characteristics to stay within a certain energy bound [134]. A method to generate a schedule, i.e. number of nodes and corresponding frequency, that minimizes the execution time of MPI programs while staying within an upper limit for energy consumption has also been presented [147]. They use a heuristic based on a novel combination of performance modeling, performance prediction, and program execution, which finds near-optimal schedules for all benchmarks used in the evaluation.

A power profiling system called PowerWatch was developed by Hotta et al., which measures and collects power information of each node in a cluster. The profiles are used to evaluate DVFS scheduling strategies that are based on phase determination [72] and task graph analysis [85], respectively.

Etinski et al. used the Paraver performance analysis tool to find load imbalances in MPI programs and developed load-balancing strategies based on DVFS to remove the slack in the unbalanced processes [44].

All aforementioned approaches regarded MPI programs but did not consider the more and more popular hybrid MPI/OpenMP applications. Li et al. developed a new algorithm for the power-aware execution of hybrid

MPI/OpenMP applications targeted specifically for large-scale systems with large shared-memory nodes [99]. The algorithm is based on DVFS and dynamic concurrency throttling (DCT), a technique to control the number of active threads that execute a portion of code, particularly in shared-memory environments. Substantial energy savings of about 4% on average could be reached with either negligible performance impact or even performance gains.

### 3.4.1 Run-Time Systems

The tools presented so far need application knowledge to reduce the energy consumption of the application. Power-aware run-time systems on the other hand try to reduce the energy consumption without application knowledge by monitoring the system behavior and prediction of future events. But, as Nils Bohr said, "Prediction is very difficult, especially about the future", that usually comes at the price of run-time increases. A suitable run-time system for HPC has to keep the run-time degradations as low as possible.

A first power-aware run-time system (PART) for high performance computing based on the $\beta$-adaptation algorithm was presented by Hsu et al. [74]. The $\beta$-adaptation algorithm is an interval-based scheduling algorithm that calculates the desired frequency of the next interval based on the MIPS (million instructions per second) rate, the allowed slowdown constraint and $\beta$ which basically is a measure of the execution time change depending on the frequency. Evaluation with the NAS benchmarks showed energy savings of up to 20% at the cost of up to 5% slowdown.

CPU MISER (Management InfraStructure for Energy Reduction), a run-time system developed by Ge et al. [55] works with an integrated performance model based on the cycles per instruction metric for performance-directed power-aware computing on generic cluster architectures. Several phases of program inactivity like memory accesses and IO are detected and exploited to save energy. Energy savings and run-time dilation are in the same range as for the PART run-time system.

Lim et al. presented a run-time system that scales the processor frequency in communication phases of MPI programs [101]. The run-time system automatically identifies such phases and sets the P-State to optimize the energy-delay-product, i.e. it allows some slowdown for higher energy savings. It can be automatically used by all MPI programs without any user involvement.

Rountree et al. developed Adagio [135], a run-time system based on critical path analysis. The critical path of a parallel application is the longest path through the program activity graph (PAG). It determines the overall run-time of the program. Adagio exploits the fact that activities that are not on the critical path can be prolonged without affecting the overall run-time by reducing the processor frequency for such activities. The critical-path detection is based solely on processor-local information, i.e. no additional communication or synchronization is needed, making Adagio very efficient. It can be used for all applications running on the machine as it generates the schedules from predicted computation times for which no application-specific knowledge is needed. Two real-world applications were used (among the NAS benchmarks) to evaluate Adagio. With less than 1% overhead, Adagio reduced the overall energy consumption by 8% and 20%, respectively.

## 3.5 Applications and Algorithms

More and more application developers care about the energy usage of their applications and work on reducing the energy consumption by investigating alternative algorithms that might require less energy or the usage of programming techniques for energy-efficiency. Most of the work is done in the field of numerical linear algebra which plays a major role in many scientific codes.

A very promising area is complexity analysis of algorithms, as an $O(n^2)$ might run faster and more efficient on a large-scale computer than an alternative algorithm with a complexity of $O(n \log n)$ because it can better utilize the hardware. However, little work has been done in this area as most developers worked hard to reduce the complexity of their algorithms.

Some work gives hints on how to write energy-efficient software by better hardware utilization, e.g. multi-threading or vectorization [149]. Other present compiler techniques to optimize the code for more energy-efficiency like loop-unrolling or recursion removal [113].

Ltaief et al. used the PowerPack framework to profile the main routines of two high performance dense linear algebra libraries, LAPACK and PLASMA, for power and energy-efficiency on a multi-core processor [107]. While LAPACK uses fork-join based thread-parallelization, PLASMA used fine-grained task parallelization with tasks operating on sub-matrices. PLASMA routines generally had a lower power consumption than the LAPACK counterparts and usually performed faster. But in general this approach helps to identify the most energy-efficient routine for a given problem.

Alonso et al. investigated energy-saving strategies for dense linear algebra algorithms, the LU factorization [4] and the QR factorization [5], implemented in a task-parallelized fashion. Two energy-saving strategies are considered, the slack reduction algorithm (SRA) and the race-to-idle algorithm (RIA), based on the analysis of the task execution graph. The slack reduction algorithm tries to reduce the slack, i.e. idle periods, in the task execution by adjusting the frequency per task via DVFS. The race to idle algorithm on the other hand tries to schedule the tasks that long idle periods can be exploited with very low power states. From a performance perspective, the RIA strategy wins in both cases, LU and QR factorization. RIA yielded in higher energy savings for the LU factorization. However, under certain circumstances the SRA strategy delivers higher energy savings for QR factorization. So it is important to choose the right strategy depending on the algorithm, the problem size, and the architecture as there is no simple general answer which policy to use.

Malkowski et al. researched the performance and energy tuning possibilities for sparse kernels like sparse matrix vector multiplication for different memory subsystems and processor power saving policies [110]. Algorithmic changes and code tuning, i.e. blocking algorithms and loop unrolling, were regarded and their performance and energy impact measured. They show that the standard implementation performs sub-optimally and argue that tuned versions should be provided in benchmarks so the most efficient for a certain platform can be chosen.

Although energy analysis of algorithms focuses on linear algebra, other areas are under investigation as well. Sundriyal et al. explored the applicability of DVFS in the quantum chemistry application GAMESS for electronic structure calculations [152]. They evaluated two ways of calculating the self consistent field method, direct and conventional. Although the direct mode is more energy-efficient, its performance is not as good if fewer cores are used.

Anzt et al. investigated the power consumption of mixed precision iterative solvers on CPUs [9] and GPUs [8] for different versions of the solvers, with and without preconditioning and using DVFS in idle states. Again the result is that the optimal configuration depends on the used hardware. However, adding a preconditioner usually improves performance and energy efficiency. Putting the processor in a sleep state while a kernel is running on the GPU further increases the energy-efficiency.

### 3.5.1 Communication Algorithms

A large portion of execution time and energy consumption of an application at large scale is due to remote data transfers, either explicitly in MPI communicating routines or implicitly in non-local memory accesses in shared-memory environments, and synchronization of processes.

Probably the highest energy-saving potential is in MPI collective operations, i.e. operations where all processes of the communicator have to participate. Usually the MPI library is only specifically tuned for massively parallel machines, while on a normal cluster a generic MPI library is used. Collective operation behavior is very implementation dependent and certainly offers room for optimization on several platforms.

All-to-All communication was investigated by Sundriyal and Sosonkina [153] while Dong et al. focused on low-power algorithms for *gather* and *scatter* operations [37]. The main idea is that collective communication is based on the exchange of point-to-point messages and not all processes participate in every step. Thus, those non-participating processes could be put to a lower power state to save energy. With an optimized communication strategy the idle phases per process can be kept as long as possible.

Similar algorithms for special types of interconnects have been studied by Kandalla et al. for InfiniBand clusters [79] and Zamani et al. for Myrinet and Quadrics QsNet [165].

Multiple MPI processes can be executed on one cluster node, as a node usually contains multi-core processors on multiple sockets which are connected via a very fast intra-node network, which is usually significantly faster than the inter-node network. With an optimized placement of MPI processes to the different nodes of a cluster high energy savings are possible [70]. A power-aware MPI task aggregation was presented by Li et al. [100]. MPI tasks that intercommunicate a lot are aggregated on the same node where communication can be replaced by a direct memory access (DMA). They proposed a framework that predicts the influence of MPI task aggregation by analyzing previously collected samples of applications. The aggregation problem, i.e. determining the optimal number of nodes and the MPI tasks per node, is formulated as a graph partitioning problem that is solved heuristically. Evaluation on systems with up to 1024 cores showed that huge energy savings with an average of over 60% are possible.

Not only MPI communication can benefit from power-aware communication algorithms, but also the PGAS (Partitioned Global Address Space) programming model, which uses implicit communication to access distant memory, can be tuned for energy-efficiency. Vishnu et al. did an extensive study on this subject [157] and presented PASCoL, the Power Aware One-Sided Communication Library. This library can detect communication slack and uses DVFS and interrupt driven execution (instead of polling) to save energy in these idle periods. The evaluation with synthetic benchmarks for the one-sided communication primitives *put*, *get*, and *accumulate* showed significant reduction in energy consumption with little performance losses, especially for large data transfers that are more bandwidth than latency dependent.

## 4. THE EECLUST PROJECT

The eeClust project[32] [116] is funded by the German Ministry for Education and Research (BMBF) with the partners University of Hamburg as project coordinator, the Jülich Supercomputing Centre (JSC) of Forschungszentrum Jülich GmbH, the Center for Information Services and High Performance Computing (ZIH) of Technische Universität Dresden, and ParTec Cluster Competence Center GmbH.

### 4.1 Project Overview

The goal of the project is the reduction of the energy consumption of applications running on HPC clusters by an integrated approach of application analysis, hardware power-state management, and system monitoring.

Figure 15 shows the overall scheme of the eeClust project [87].

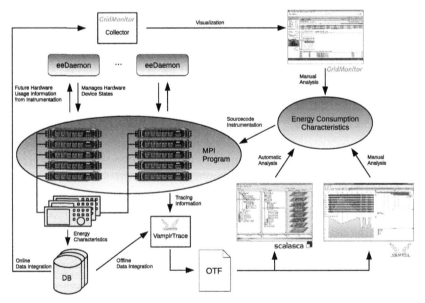

**Fig. 15.** eeClust Project Chart.

[32] http://www.eeclust.de/index.en.html

**Table 2** Hardware eeClust cluster.

| Component | Intel Nodes | AMD Nodes |
|---|---|---|
| Processor | Intel Xeon EP QC X5560 [29, 30] | AMD Opteron 6168 [1] |
| Memory | 12 GB (6x2) DDR3 1333 MHz | 32 GB (8X4) DDR3 1333 MHz |
| Network | Intel 82574L [31] | Intel 82576 [32] |
| HDD | Seagate Barracuda 7200.12 [143] | Seagate Barracuda 7200.12 [143] |
| Power supply | 80 + Gold | 80 + Gold |

As no production machine at any of the centers involved in the project was capable of fine-grained power measurements we procured a test cluster which should reflect the latest x86 technology. So we got a heterogenous 10-node cluster where five nodes are equipped with two Intel Xeon processors and five nodes with two AMD Opteron processors. Table 2 lists the detailed hardware configuration.

All cluster nodes are connected to three four-channel LMG450 power meters from ZES Zimmer [59]. These power meters offer high frequency measurement of up to 20 Hz at a very high accuracy of 0.1%. Power measurement is enabled all the time and measurement values are stored in a database. This allows the monitoring component to display historic values and enables the integration into the measurement library.

We obtain application traces with the VampirTrace measurement library and correlate the behavior of the parallel programs to their power consumption using the performance analysis tools Vampir [124] and Scalasca [162]. This information is used to identify phases in which certain hardware is not utilized and could be put into a lower power state to save energy, as presented in Section 4.2.

With the knowledge of the applications energy characteristic we can instrument the application to express the hardware requirements to a daemon process that switches the hardware power-states. The API and the daemon are discussed in Section 4.3.

As cluster middle-ware we used ParTec's ParaStation,[33] which provides cluster management features, a process manager, and an efficient MPI

---

[33] http://www.par-tec. com/products/parastationv5.html

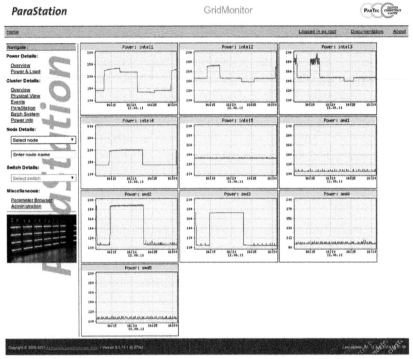

**Fig. 16.** Screenshot of the parastation gridmonitor cluster monitoring software. It shows the ten nodes of the eeClust-cluster in the newly developed power monitoring view.

implementation. The monitoring component ParaStation GridMonitor[34] has been extended in the project by a power consumption view as shown in Fig. 16.

## 4.2 Application Analysis

Many HPC Application developers have—at least to some extent—experience with performance analysis tools—not so many have much experience with power and energy profiling and many have no intention of learning yet another tool. To build upon the already existing knowledge of application developers, we decided to extend our well-known and successful performance analysis tools Vampir [89] and Scalasca [57] to do a combined

---

[34] http://w ww.par-tec.com/products/parastation-gridmonitor.html

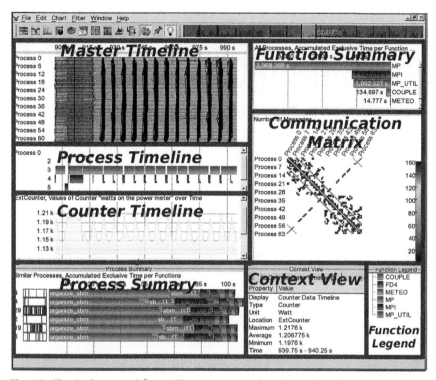

**Fig. 17.** The Scalasca workflow—The gray rectangles denote (parallel) programs, the rectangles with the kinked corner symbolize (multiple) files. White Rectangles denote third party tools. The left part (a) demonstrates the general procedure for application tracing and could also be applied for VampirTrace. The right part (b) shows the workflow for the automatic analysis which is distinct for Scalasca.

performance and energy analysis. And in fact performance tuning should always be the first step in the energy tuning of an application, as a shorter runtime in general corresponds to less energy consumption, although the average power consumption may increase due to a higher utilization of components.

The workflow for application tracing and analysis is outlined in Fig. 17.

The left side (1) shows basically the general workflow for tracing and is also valid for VampirTrace. The source code is instrumented to insert calls to the measurement library at every function enter and exit—that is usually done by using compiler hooks, but also user instrumentation is possible. When executing the instrumented executable, every process writes a trace file where every event—function enter/exit, MPI messages, etc.—is

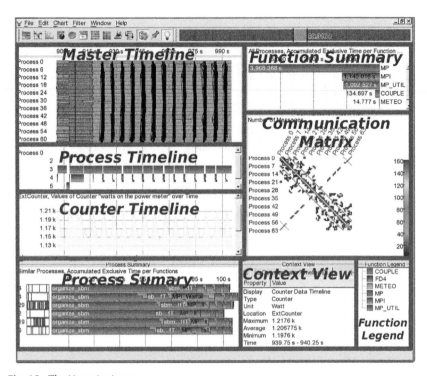

**Fig. 18.** The Vampir views.

recorded with the corresponding timestamp, so the whole execution of the application can be reconstructed afterward. Additionally, hardware counters can be recorded for every event so a fine-grained understanding of the application behavior is possible.

Vampir, developed at the ZIH in Dresden, is a performance analysis tool for a manual analysis of trace files. It parses the files and presents the results in several possible views as outlined in Fig. 18.

Vampir provides a set of timelines—a master timeline which shows the behavior of all processes and their interaction, i.e. the communication between the processes, a counter timeline to visualize hardware counter metrics, and a process timeline for an in-depth study of the threads in a multi-threaded program. Further possible views contain a communication matrix, process and functions summaries, a context view, and a legend.

Originally VampirTrace was only able to collect hardware counters from a small number of sources, primarily from the PAPI library. But the Dresden folks developed a plug-in infrastructure to obtain counters from arbitrary sources and include the values in a trace file [140]. With the plug-in counter infrastructure, different types of counters are possible, distinguished by whether the counter value is absolute—like a sensor reading—or accumulated—like the traditional hardware performance counters like cache-misses—and when the counters are recorded. The following four types of counters exist:

- *Synchronous pluginsi*—The counter values are recorded every time VampirTrace generates an event. This corresponds to the traditional method of obtaining hardware counters.
- *Asynchronous plugins*—The plugin collects the data by itself and VampirTrace asks it to report the counters any time it generates an event.
- *Asynchronous callback plugins*—The plugin collects the data by itself and reports the data to VampirTrace using a callback function.
- *Asynchronous post-mortem plugins*—The plugin collects the data by itself and reports the data to VampirTrace at the end of the tracing.

In the project an asynchronous post-mortem plugin was developed to query the database for the power data of the nodes and merge them into the trace files. Using the post-mortem approach minimizes measurement overhead as all communication is done at the end of the execution of the application. An example screenshot of Vampir showing power data is given in Fig. 21. This plugin enables the correlation of the power consumption to the applications behavior by mapping the power data with data from other counters that are recorded concurrently.

Scalasca is jointly developed by JSC and the German Research School for Simulation Sciences in Aachen. Scalasca's distinctive feature is an automatic analysis of the trace files for patterns indicating workload imbalances as illustrated in the right part (2) of Fig. 17. The analyzer, which is a parallel program of its own, has to be run with as many processes as the original application. It "replays" the communication of the traced application, but instead of the original data it transfers timestamps and other data used to recognize and quantify wait-states, i.e. phases where one process had to wait for another process to finish before it can proceed, so no progress is possible on that processor. Wait-states are usually caused by unequal workload distribution due to bad load balancing and thus indicate a performance degradation. However, sometimes wait-states are unavoidable due to data dependencies among processes.

Some exemplary patterns that Scalasca is able to find are shown in Fig. 19. Scalasca is able to detect wait-states in point-to-point communication

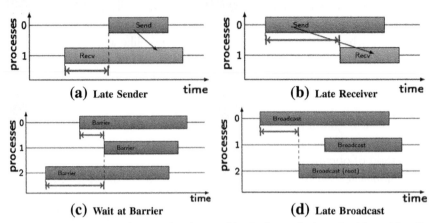

**Fig. 19.** Some patterns that can be detected by Scalasca. Scalasca is able to identify wait-states for point-to-point (a,b) as well as for collective communication (c,d). The time indicated by the double arrows is considered waiting time.

(Fig. 19a and b) as well as in collective operations (Fig. 19c and d). Scalasca is able to determine wait-states for basically all MPI functions, even one–sided communication is supported. There is further support for some OpenMP patterns.

The analyzer creates an analysis report containing all metrics that can be shown in the Cube analysis report browser as illustrated in Fig. 20. Cube offers a three panel layout showing the metric tree in the left panel, the distribution of the selected metric on the call-path in the middle panel, and the distribution of the selected metric on the selected call-path in the right panel.

Scalasca supports not only tracing but also exact call-path based profiling of an application. In contrast to tracing, no events are stored which reduces the measurement overhead significantly. The recorded metrics like time, visits of a function, etc., are aggregated over the run-time of the application. The resulting profile report has the same structure as the trace analysis report and can also be displayed by the Cube browser.

Usually the profile report is used to identify function that can be "filtered" during tracing, i.e. no events are recorded for that functions to reduce measurement overhead. This is especially important for the detection of the wait-states as run-time dilation can hide existing wait-states or even induce new wait-states which do not occur in the uninstrumented application.

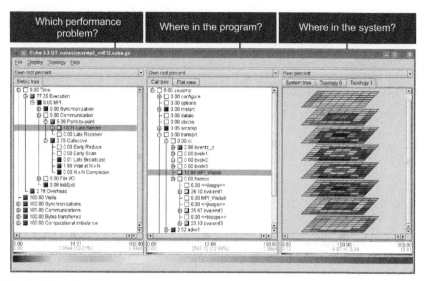

**Fig. 20.** The cube performance analysis report browser.

### 4.2.1 Automatic Energy Analysis

Up to now Scalasca focused solely on performance problems but, unfortunately, wait-states cannot always be prevented—be it due to sequential parts in a program or insufficient parallelism. So the question arises what the wait-states mean in terms of energy consumption [88].

Figure 21 shows a Vampir screenshot for an artificially created wait-at-barrier pattern on an Intel Xeon node with eight processes. After an unequally distributed calculation phase the processes sleep in a manner that they enter the MPI_Barrier one after the other (1). Obviously the power consumption rises as more processes enter the barrier.

This is due to an implementation detail in the MPI library. Most MPI libraries do a so-called busy-waiting, i.e. a process that waits for progress on another process polls at the highest frequency to immediately notice when the other process is ready so it can continue work. While this is very good from a performance perspective, this behavior wastes a lot of energy. Many MPI libraries can switch to an interrupt-based notification, i.e. the waiting process does an idle-waiting and is notified by an interrupt so it does not have to poll for progress. However, this can have severe performance implications.

We examined how much energy could be saved by switching to a lower power state during a wait-state.

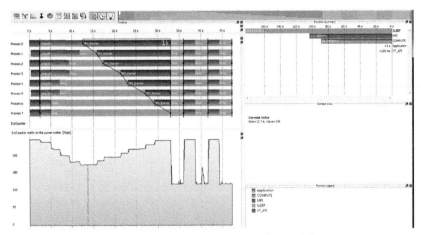

**Fig. 21.** Screenshot of Vampir showing the master timeline and the power consumption in the counter timeline. One can see that the more processes enter the MPI Barrier (a) the higher the power consumption of the node. The displayed pattern corresponds with the Scalasca Wait at Barrier pattern (Fig. 19c).

The energy-saving potential $ESP$ for idle-waiting, which poses an upper limit for the saving potential, can be calculated as shown in Eqn (8):

$$\text{ESP} = \max_{p \in PS} \left( (t_w * A_{p_0}) - \left(t_w - t_{T_{p,p_0}}\right) * I_p + E_{T_{p,p_0}} \right). \tag{8}$$

Here $PS$ is a set of power states, $p \in PS$ is a power state, $t_w$ is the waiting time, $t_{T_{p,p_0}}$ is the time, and $E_{T_{p,p_0}}$ the energy needed for the transition from power state $p_0$ to power state $p$ and back. $A_p$ is the active power consumption and $I_p$ the idle power consumption in power state $p$.

The term $t_w - A_{p_0}$ is the energy actually spent in the wait-state, $(t_w - t_{T_{p,p_0}}) * I_p$ is the energy spent in idle-waiting without considering the transition time which is already covered in the transition energy $E_{T_{p,p_0}}$.

With that we can calculate the more realistic energy-saving potential while busy-waiting in a lower power state $ESP\_BW$ as in Eqn (9):

$$ESP\_BW = \max_{p \in PS} ((t_w * A_{p_0}) - (t_w - t_{T_{p,p_0}}) * A_p + E_{T_{p,p_0}}). \tag{9}$$

Additionally, in each case the power state which yields the highest energy savings is stored and can be displayed as a metric in Cube.

Figure 22 gives an example of the analysis of PEPC [161], a plasma physics Barnes-Hut tree code developed at JSC, on JSC's Juropa cluster using 1024

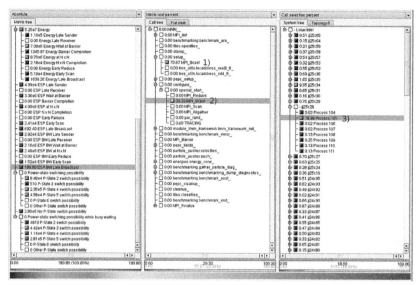

**Fig. 22.** Screenshot of the energy analysis of PEPC on the Juropa cluster with 1024 processes. It shows the energy-saving potential in Late Broadcasts.

processes. It shows the energy-saving potential in Late Broadcasts. There are two broadcasts in the application, with 70% of the energy-saving potential in broadcast (1) and 30% in broadcast (2). 19% of the saving potential of the second broadcast is on process 105 (3) while the energy-saving potential on the other processes is relatively small.

## 4.3 Hardware Management

The analysis results are used to control the hardware power-states to turn off unused components at run-time.

Traditional hardware power management (see Fig. 23a), the operating system switches components to a lower power states after a certain time of inactivity and back if the component is needed again. This is not really efficient for two reasons: first, it wastes energy in the idle time before switching the power states and second, it induces a performance penalty because the component is only switched back when it is already needed. Thus, the traditional hardware power management is not really suited for HPC.

For eeClust we developed a new application-aware hardware management [114] as displayed in Fig. 23b. The heart of this approach is the eeDaemon, a system process that manages the hardware power states.

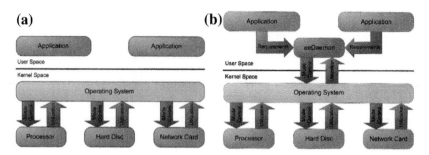

**Fig. 23.** Hardware power state management.

It currently supports three different devices, CPU, NIC, and HDD with three power states each, a high, a medium, and a low-power state, which can be differently implemented for each device. The eeDaemon provides a simple application programming interface (API) to manually instrument the application and enable processes to communicate its hardware requirements, so hardware can be put to lower power states in phases where it is not utilized. The API consists of the following five functions:

- `ee_init`—Registers the process at the eeDaemon.
- `ee_finalize`—Unregisters the process at the eeDaemon.
- `ee_dev_state`—Request the daemon to set a device to a certain power state.
- `ee_ready_in`—Notify the daemon that a device is needed in the specified time.
- `ee_idle_for`—Notify the daemon that a device is not needed for the specified time.

An outstanding feature of the eeDaemon is that it is aware of shared resources, e.g. the network interface is usually shared by all processors on a node. The daemon always chooses the maximum requested power state, i.e. if three of four processes do not need the NIC and the fourth process needs it, it will be put in the highest power state. Only when the fourth process does not need it any more it is put to a lower power state. To know how many processes are running on a node, the eeDaemon is tightly integrated in the resource manager. If an uninstrumented job is running on a node, always the maximum power states are used, even if it shares the node with an instrumented application.

```
1   int main() {
2   ...
3     for(i=0; i<100; i++) {
4        do_work(); //app. 4s
5        communicate_results(); // app. 2s
6        write_checkpoint(); // app. 3s
7     }
8   ...
9   }
```

**Listing 1.** Original code.

```
1   int main() {
2   ee_init(); // Register process at eeDaemon
3   ee_dev_state(DEV_CPU,STATE_MED);
4   ee_dev_state(DEV_NET,STATE_UNUSED);
5   ee_dev_state(DEV_HDD,STATE_UNUSED);
6   ...
7     ee_dev_state(DEV_CPU,STATE_MAX);
8     for(i=0; i<100; i++) {
9        ee_dev_ready_in(DEV_NET,4);
10       ee_dev_ready_in(DEV_HDD,6);
11       do_work(); //app. 4s
12       ee_dev_ready_in(DEV_CPU,5);
13       communicate_results(); // app. 2s
14       ee_dev_idle_for(DEV_NET,7);
15       write_checkpoint(); // app. 3s
16    }
17  ...
18  ee_finalize();
19  }
```

**Listing 2.** Instrumented code.

Listings 1 and 2 give an example of how the instrumentation might look like for a typical loop in a scientific code. It performs some computation (line 4), communicates the results (line 5) and then writes a checkpoint (line 6). The times for each of these phases have been obtained by a former performance analysis.

We evaluated the approach with two applications, a solver for partial differential equations called partdiff-par and the aforementioned PEPC application. Multiple experiments have been performed, running with maximum and minimum frequency and with eeDaemon instrumentation.

**Table 3** Evaluation partdiff-par Intel nodes.

| Setup | TTS (s) | ETS (J) | Slowdown (%) | Energy-Saving (%) |
|---|---|---|---|---|
| CPU Max | 260.2 | 130479.6 | – | – |
| CPU Min | 424.5 | 159110.3 | 63.16 | −21.94 |
| OnDemand | 279.2 | 136561.3 | 7.30 | −4.66 |
| Instrumented | 271.3 | 124099.9 | 4.27 | 4.89 |

**Table 4** Evaluation PEPC AMD nodes.

| Setup | TTS (s) | ETS (J) | Slowdown (%) | Energy-Saving (%) |
|---|---|---|---|---|
| CPU Max | 950.02 | 1197495.75 | – | – |
| CPU Min | 2016.71 | 1759821.48 | 112.28 | −46.95 |
| sum_force | 967.12 | 1153188.4 | 1.8 | 3.7 |
| global | 990.87 | 1119658.52 | 4.3 | 6.5 |

Table 3 shows the results for partdiff-par on the Intel Xeon nodes of the cluster. Partdiff-par was configured with a checkpoint between two calculation phases and the matrix size was chosen to fill the memory. Both runs with minimum frequency and with the ondemand governor of the operating system yielded in higher energy consumption due to a longer run-time. In the instrumented version the processors are set to the lowest power during check-pointing and network and disks are turned down in the calculation phases. For this small example that yielded in nearly 5% energy savings at a slowdown of slightly more than 4% (see Table 4).

PEPC was evaluated on the AMD Opteron nodes of the cluster simulating 512,000 particles in 50 time-steps with two different instrumentations. Notable in the PEPC case is the very bad result when running at the lowest frequency. Due to the highly increased run-time, which was more than twice of the original run-time, nearly 50% more energy was consumed. The sum_force instrumentation is a fairly simple instrumentation where only for the most computational intensive routine (the sum_force routine) the processors ran at full speed. For the global instrumentation also network and disk were considered and switched before the communication and I/O-phases, respectively. The instrumented versions led to energy savings of 3.7% and 6.5% and a performance degradation of 1.8% and 4.3%, respectively.

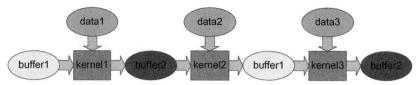

**Fig. 24.** The kernel sequence of the eeMark benchmark. Each kernel uses three buffers, an input, an output, and a data buffer. The output buffer of a kernel serves as input buffer for the subsequent kernel. The data buffer is predefined for each kernel.

## 4.4 Benchmarking

The SPECpower benchmark[35] is the first benchmark that evaluates power and performance of servers [95]. However, the SPECpower is not suited for HPC as it measures an enterprise focused server side Java workload, so the results are not representative for HPC applications.

In the eeClust project a benchmark for both power and performance efficiency has been developed with special focus on HPC-tailored workloads. This benchmark, called eeMark [120], is used to verify the usability of the eeClust approach of application instrumentation and the results can serve as input for further analysis.

HPC-tailored workloads are generated by running multiple kernels that stress certain system components. The eeMark provides compute kernels that generate high load on CPUs and/or memory, MPI-based communication kernels that stress the network, and I/O kernels that put a high load on the file system. Kernel sequences generate phases with varying utilization of components, which reflect the typical phases in HPC applications. Each kernel uses three buffers, an input buffer, an output buffer, and a data buffer. The output buffer of one kernel serves as input buffer for the subsequent kernel, the data buffer is specific for each kernel. The sequence of kernels and the input data can be chosen manually, which allows the workload scale easily with the system size, resulting in comparable results across a range of platforms. Figure 24 illustrates that workflow.

The benchmark is not tied to specific power measurement devices, it rather provides an interface to implement a custom measurement method. However, it already provides an interface for the SpecPower daemon (ptd), so systems that were benchmarked with SpecPower are supported out of the box.

Upon completion, the benchmark determines a performance and an efficiency rating. For that, all kernels report the number of performed operations and these operations are weighted according to their effort. The performance

[35] http://www.spec.org/power_ssj200 8/

rating then is the number of weighted operations per second, the efficiency rating is the number of weighted operations per watt.

As the benchmark is highly configurable it can be used to determine the strengths and weaknesses in efficiency of a platform for certain operations. Having this information for several platforms and an application profile it is easy to find the suitable platform for the application. That information, correlated with hardware counter values, can also help the analysis tools to better guide the application developer to find energy-tuning possibilities.

## 5. CONCLUSION

In recent years, power consumption have become a major topic in HPC for both operational and environmental perspectives. Additionally, the power consumption of the hardware and the energy-efficiency of software are major design constraints in the development of future supercomputers in the race to Exascale.

The hardware vendors have integrated a great number of power management features in current generation of components, especially CPUs, which we presented here. We outlined the current developments for more power efficient hardware that might become the base of tomorrow's supercomputers. We also presented several special HPC architectures that tackle the power consumption problem with different approaches, leading to the most efficient supercomputers today.

Software solutions are necessary for an efficient operation of supercomputers at all levels, from the systems software over the running application to the cluster management software. The hardware vendors present more and more sophisticated tools to manage hardware at several levels from systems to components. We gave an overview of the current research in scheduling and tools for energy-efficent HPC. Further the first attempts for algorithm analysis have been presented.

Finally, we presented the eeClust project in detail, which offers a holistic software set for energy-aware HPC consisting of application analysis using well-known performance analysis tools that have been extended for energy analysis, a run-time system for efficient hardware management with an application instrumentation interface, and a monitoring component. The eeMark benchmark is the first portable HPC-tailored power-efficiency benchmark.

Yet there is still a lot of work to do to reach the 50 Gflops/Watt goal for an Exascale system in the time frame of 2018–2020, both from a hardware and from a software perspective, with many enthusiastic teams in academics and industry working on these challenges. Probably the major challenge is to increase the awareness of application developers for the energy consumption and energy-efficiency of their applications. Although not involved with the operational problems of running a large-scale system, their codes determine the energy-efficiency of the machine. We hope that this chapter can contribute to that.

## ACKNOWLEDGMENTS

The eeClust project was funded by the German Federal Ministry of Education and Science (BMBF—Bundesministerium für Bildung und Forschung) under grant 01—H08008E within the call "HPC-Software für skalierbare Parallelrechner."

The author thanks all the project partners from the eeClust project, especially Timo Minartz from the University of Hamburg, who did most of the work on the API and eeDaemon and who administrated the cluster, Daniel Molka from TU Dresden, who developed the eeMark benchmark and is responsible for the Vampir measurements, and Stephan Krempel from the ParTec Cluster Competence Center GmbH who developed the new power monitoring view for the GridMonitor. Further thanks go to Professor Thomas Ludwig from University of Hamburg and the German Climate Computing Centre, who coordinated the project and started and chaired the International Conference on Energy-Aware High Performance Computing (EnA-HPC).

I want to thank the whole Scalasca development team at Juelich Supercomputing Centre and GRS Aachen, especially Bernd Mohr, Markus Geimer, and Brian Wylie.

Finally I'd like to thank the team of the energy-efficiency working group of the Exascale Innovation Centre (EIC), a collaboration of IBM Research and Development Germany and the Juelich Supercomputing Centre, namely Willi Homberg and Professor Dirk Pleiter from JSC as well as Hans Boettinger and Michael Hennecke from IBM.

## LIST OF ABBREVIATIONS

| | |
|---|---|
| ACPI | Advanced Configuration and Power Interface |
| AEM | Active Energy Manager |
| API | Application Programming Interface |
| ASIC | Application-Specific Integrated Circuit |
| ATU | Address Translation Unit |

| CPU | Central Procesing Unit |
| DCiE | Data Center Infrastructure Efficiency |
| DCT | Dynamic Concurrency Throttling |
| DEEP | Dynamic Exascale-Entry Platform |
| DMA | Direct Memory Access |
| DVFS | Dynamic Voltage and Frequency Scaling |
| DVS | Dynamic Voltage Scaling |
| EDP | Energy-Delay-Product |
| EEHPCWG | Energy-Efficient High Performance Computing Working Group |
| FIFO | First In, First Out |
| FLOP | Floating-Point OPeration |
| FPGA | Field-Programmable Gate Array |
| GPGPU | General Purpose Graphics Processing Unit |
| GPU | Graphics Processing Unit |
| HDD | Hard Disk Drive |
| HMC | Hybrid Memory Cube |
| HPC | High Performance Computing |
| HPL | High Performance Linpack |
| IOPS | Input/Output Operations Per Second |
| J | Joule |
| JSC | Juelich Supercomputing Centre |
| LLNL | Lawrence Livermore National Lab |
| MIC | Many Integrated Cores |
| MIPS | Million Instructions Per Second |
| MPI | Message passing Interface |
| NCSA | National Center for Supercomputing Applications |
| NoC | Network-on-Chip |
| NTC | Near-Threshold Computing |
| PAG | Program Activity Graph |
| PAPI | Performance Application Programming Interface |
| PCU | Package Control Unit |
| PCM | Phase-Change Memory |
| PDU | Power Distribution Unit |
| PGAS | Partitioned Global Address Space |
| PPE | Power Processing Element |
| PUE | Power Usage Effectiveness |
| QCD | Quantum ChromoDynamics |
| RAM | Random Access Memory |
| RDMA | Remote Direct Memory Access |
| RMA | Remote Memory Access |
| SPE | Synergistic Processing Element |
| SPEC | Standard Performance Evaluation Corporation |
| SMT | Simultaneous MultiThreading |
| SoC | System-on-a-Chip |
| SSD | Solid State Drive |
| TCO | Total Cost of Ownership |
| TLB | Translation Lookaside Buffer |

TPMD        Thermal and Power Management Device
VELO        Virtualized Engine for Low Overhead
W           Watt

# REFERENCES

[1] Advanced micro devices, Inc., Family 10h AMD opteronTM processor product data sheet, June 2010.

[2] Phil Ainsworth, Miguel Echenique, Bob Padzieski, Claudio Villalobos, Paul Walters, Debbie Landon, Going green with IBM systems director active energy manager, IBM.

[3] Hadil Al-Daoud, Issam Al-Azzoni, Douglas G. Down, Power-aware linear programming based scheduling for heterogeneous computer clusters, Future Generation Computer Systems 28 (5) (2012) 745–754. Special Section: energy efficiency in large-scale distributed systems.

[4] Pedro Alonso, Manuel Dolz, Francisco Igual, Rafael Mayo, Enrique Quintana-Ortí, Dvfs-control techniques for dense linear algebra operations on multi-core processors, Computer Science—Research and Development pages 1–10, doi: 10.1007/s00450-011-0188-7.

[5] Pedro Alonso, Manuel Dolz, Rafael Mayo, Enrique Quintana-Ortí, Energy-efficient execution of dense linear algebra algorithms on multi-core processors, Cluster Computing 1–13, doi: 10.1007/s10586-012-0215-x.

[6] Ed Anderson, Jeff Brooks, Charles Grassl, Steve Scott, Performance of the cray T3E multiprocessor, in: Proceedings of the 1997 ACM/IEEE Conference on Supercomputing (CDROM), Supercomputing '97, ACM, New York, NY, USA, 1997, 1–17.

[7] Kumar Anshumali, Terry Chappel, Wilfred Gomes, Jeff Miller, Nasser Kurd, Rajesh Kumar, Circuit and process innovations to enable high performance, and power and area efficiency on the nehalem and westmere family of intel processors, Intel Technology Journal 14 (3) 2010 104–127.

[8] Hartwig Anzt, Maribel Castillo, Juan Fernández, Vincent Heuveline, Francisco Igual, Rafael Mayo, Enrique Quintana-Ortí, Optimization of power consumption in the iterative solution of sparse linear systems on graphics processors, Computer Science—Research and Development 1–9, doi: 10.1007/s00450-011-0195-8.

[9] Hartwig Anzt, Björn Rocker, Vincent Heuveline, Energy efficiency of mixed precision iterative refinement methods using hybrid hardware platforms, Computer Science—Research and Development 25 (2010) 141–148, doi: 10.1007/s00450-010-0124-2.

[10] Axel Auweter, Arndt Bode, Matthias Brehm, Herbert Huber, Dieter Kranzlmüller. Principles of energy efficiency in high performance computing, in: Dieter Kranzlmüller, A Toja (Eds.), Information and Communication on Technology for the Fight against Global Warming, vol. 6868 Lecture Notes in Computer Science, Springer Berlin/Heidelberg, 2011, pp. 18–25, doi: 10.1007/978-3-642-23447-7_3.

[11] H. Baier, H. Boettiger, M. Drochner, N. Eicker, U. Fischer, Z. Fodor, A. Frommer, C. Gomez, G. Goldrian, S. Heybrock, D. Hierl, M. Hüsken, T. Huth, B. Krill, J. Lauritsen, T. Lippert, T. Maurer, B. Mendl, N. Meyer, A. Nobile, I. Ouda, M. Pivanti, D. Pleiter, M. Ries, A. Schäfer, H. Schick, F. Schifano, H. Simma, S. Solbrig, T. Streuer, K.-H. Sulanke, R. Tripiccione, J.-S. Vogt, T. Wettig, and F. Winter. Qpace: power-efficient parallel architecture based on IBM powerxcell 8i, Computer science—research and development 25 (2010) 149–154, doi: 10.1007/s00450-010-0122-4.

[12] D.H. Bailey, E. Barszcz, J.T. Barton, D.S. Browning, R.L. Carter, L. Dagum, R.A. Fatoohi, P.O. Frederickson, T.A. Lasinski, R.S. Schreiber, H.D. Simon, V. Venkatakrishnan, S.K. Weeratunga, The nas parallel benchmarks summary and preliminary results, in: Proceedings of the 1991 ACM/IEEE Conference on Supercomputing 1991, Supercomputing '91, 1991, pp. 158–165.

[13] L.A. Barroso, U. Hölzle, The case for energy-proportional computing, Computer 40 (12) (2007) 33–37.

[14] Robert Basmadjian, Hermann de Meer, Evaluating and modeling power consumption of multi-core processors, in: 2012 Third International Conference on Future Energy Systems: Where Energy, Computing and Communication Meet (e-Energy), 2012, pp. 1–10.

[15] C. Batten, A. Joshi, J. Orcutt, A. Khilo, B. Moss, C.W. Holzwarth, M.A. Popovic, Hanqing Li, H.I. Smith, J.L. Hoyt, F.X. Kartner, R.J. Ram, V. Stojanovic, K. Asanovic, Building many-core processor-to-dram networks with monolithic CMOS silicon photonics, IEEE Micro 29 (4) (2009) 8–21.

[16] Costas Bekas, Alessandro Curioni, A new energy aware performance metric, Computer Science—Research and Development 25 (2010) 187–195, doi: 10.1007/s00450-010-0119-z.

[17] Keren Bergman, Shekhar Borkar, Dan Campbell, William Carlson, William Dally, Monty Denneau, Paul Franzon, William Harrod, Jon Hiller, Sherman Karp, Stephen Keckler, Dean Klein, Robert Lucas, Mark Richards, Al Scarpelli, Steven Scott, Allan Snavely, Thomas Sterling, R. Stanley Williams, Katherine Yelick, Keren Bergman, Shekhar Borkar, Dan Campbell, William Carlson, William Dally, Monty Denneau, Paul Franzon, William Harrod, Jon Hiller, Stephen Keckler, Dean Klein, Peter Kogge, R. Stanley Williams, Katherine Yelick, Exascale Computing Study: Technology Challenges in Achieving Exascale Systems, 2008.

[18] Rudolf Berrendorf, Heribert C. Burg, Ulrich Detert, Rüdiger Esser, Michael Gerndt, Renate Knecht, Intel Paragon xp/s—Architecture, Software Environment, and Performance, 1994.

[19] R.A. Bheda, J.A. Poovey, J.G. Beu, T.M. Conte, Energy efficient phase change memory based main memory for future high performance systems, in: International Green Computing Conference and Workshops (IGCC) 2011, 2011, pp. 1–8.

[20] Christian Bischof, Dieter an Mey, Christian Iwainsky, Brainware for green HPC, Computer Science—Research and Development 1–7, doi: 10.1007/s00450-011-0198-5.

[21] OpenMP Architecture Review Board, Openmp application program interface, version 3.1, July 2011, available at: http://openmp.org.

[22] Luigi Brochard, Raj Panda, Sid Vemuganti, Optimizing performance and energy of hpc applications on POWER7, Computer Science Research and Development 25 (3–4) (2010) 135–140.

[23] Martha Broyles, Chris Francois, Andrew Geissler, Michael Hollinger, Todd Rosedahl, Guillermo Silva, Jeff Van Heuklon, Brian Veale. IBM energyscale for POWER7 processor-based systems, Technical report, IBM, February 2010.

[24] T. Budnik, B. Knudson, M. Megerian, S. Miller, M. Mundy, W. Stockdell, Blue Gene/Q resource management architecture, in: IEEE Workshop on Many-Task Computing on Grids and Supercomputers (MTAGS) 2010, November 2010, pp. 1–5.

[25] Van Bui, Boyana Norris, Kevin Huck, Lois Curfman McInnes, Li Li, Oscar Hernandez, Barbara Chapman, A component infrastructure for performance and power modeling of parallel scientific applications, in: Proceedings of the 2008 compFrame/HPC-GECO workshop on Component based high performance, CBHPC '08, ACM, New York, NY, USA, 2008, pp. 6:1–6:11.

[26] Dong Chen, N.A. Eisley, P. Heidelberger, R.M. Senger, Y. Sugawara, S. Kumar, V. Salapura, D.L. Satterfield, B. Steinmacher-Burow, J.J. Parker, The IBM blue Gene/Q interconnection network and message unit, in: International Conference for High Performance Computing, Networking, Storage and Analysis (SC), 2011, November 2011, pp. 1–10.

[27] L. Chua, Memristor-the missing circuit element, IEEE Transactions on Circuit Theory 18 (5) (1971) 507–519.

[28] Wu chun Feng, K.W. Cameron, The green500 list: encouraging sustainable super-computing, Computer 40 (12) (2007) 50–55.

[29] Intel Corporation, Intel® Xeon® Processor 5500 Series Datasheet Volume 1, March 2009.

[30] Intel Corporation, Intel® Xeon® Processor 5500 Series Datasheet Volume 2, April 2009.

[31] Intel Corporation, Intel® 82574 GbE Controller Family, February 2010.

[32] Intel Corporation, Intel® 82576 Gigabit Ethernet Controller Datasheet, June 2010.

[33] William M. Corwin, Douglass C. Locke, Karen D. Gordon, Overview of the ieee posix p1003.4 realtime extension to Posix, IEEE Real-Time System Newsletter 6 (1) (1990) 9–18.

[34] Kim Cupps, Mary Zosel (Eds.), The 4th Workshop on HPC Best Practices: Power Management—Workshop Report, March 2011.

[35] Howard David, Chris Fallin, Eugene Gorbatov, Ulf R. Hanebutte, Onur Mutlu, Memory power management via dynamic voltage/frequency scaling, in: Proceedings of the 8th ACM international conference on Autonomic computing, ICAC '11, ACM, New York, NY, USA, 2011, pp. 31–40.

[36] V. Delaluz, M. Kandemir, N. Vijaykrishnan, M.J. Irwin, Energy-oriented compiler optimizations for partitioned memory architectures, in: Proceedings of the 2000 international conference on Compilers, architecture, and synthesis for embedded systems, CASES '00, ACM, New York, NY, USA, 2000, pp. 138–147.

[37] Yong Dong, Juan Chen, Xuejun Yang, Canqun Yang, Lin Peng, Low power optimiza-tion for MPI collective operations, in: The 9th International Conference for Young Computer Scientists, 2008, ICYCS 2008, November 2008, pp. 1047–1052.

[38] Jack J. Dongarra, Piotr Luszczek, Antoine Petitet, The linpack benchmark: past, present and future, Concurrency and Computation: Practice and Experience 15 (9) (2003) 803–820.

[39] David Donofrio, Leonid Oliker, John Shalf, Michael F. Wehner, Chris Rowen, Jens Krueger, Shoaib Kamil, Marghoob Mohiyuddin, Energy-efficient computing for extreme-scale science, Computer 42(11) (2009) 62–71.

[40] R.G. Dreslinski, M. Wieckowski, D. Blaauw, D. Sylvester, T. Mudge, Near-threshold computing: reclaiming moore's law through energy efficient integrated circuits, Proceedings of the IEEE 98 (2) (2010) 253–266.

[41] N. Eicker, T. Lippert, An accelerated cluster-architecture for the exascale, in: PARS '11, PARS-Mitteilungen, Mitteilungen—Gesellschaft für Informatik e.V., Parallel-Algorithmen und Rechnerstrukturen, number 28, October 2011, pp. 110–119.

[42] J. Enos, C. Steffen, J. Fullop, M. Showerman, Guochun Shi, K. Esler, V. Kindratenko, J.E. Stone, J.C. Phillips, Quantifying the impact of gpus on performance and energy efficiency in HPC clusters, in: Green Computing Conference, 2010 International, August 2010, pp. 317–324.

[43] M. Etinski, J. Corbalan, J. Labarta, M. Valero, Optimizing job performance under a given power constraint in hpc centers, in: Green Computing Conference, 2010 International, August 2010, pp. 257–267.

[44] M. Etinski, J. Corbalan, J. Labarta, M. Valero, A. Veidenbaum, Power-aware load balancing of large scale MPI applications, in: IEEE International Symposium on Parallel Distributed Processing, 2009, IPDPS 2009, May 2009, pp. 1–8.

[45] Maja Etinski, Julita Corbalan, Jesus Labarta, Mateo Valero, Utilization driven power-aware parallel job scheduling, Computer Science—Research and Development 25, doi: 207–216, 2010. 10.1007/s00450-010-0129-x.

[46] Xiaobo Fan, Carla Ellis, Alvin Lebeck, The synergy between power-aware memory systems and processor voltage scaling, in: Babak Falsafi, T. VijayKumar (Eds.),

Power-Aware Computer Systems, vol. 3164 of Lecture Notes in Computer Science, Springer Berlin/Heidelberg, 2005, pp. 151–166, doi: 10.1007/978-3-540-28641-7_12.

[47] Zhe Fan, Feng Qiu, Arie Kaufman, Suzanne Yoakum-Stover, Gpu cluster for high performance computing, in: Proceedings of the 2004 ACM/IEEE conference on Supercomputing, SC '04, IEEE Computer Society, Washington, DC, USA, 2004, p. 47.

[48] Dietmar Fey, Max Schneider, Jürgen Jahns, Hans Knuppertz, Optical multiplexing techniques for photonic clos networks in high performance computing architectures, in: Shlomi Dolev, Mihai Oltean (Eds.), Optical Super Computing, volume 5882 of Lecture Notes in Computer Science, Springer Berlin/Heidelberg, 2009, pp. 110–123, doi: 10.1007/978-3-642-10442-8_15.

[49] M. Floyd, M. Allen-Ware, K. Rajamani, B. Brock, C. Lefurgy, A.J. Drake, L. Pesantez, T. Gloekler, J.A. Tierno, P. Bose, A. Buyuktosunoglu, Introducing the adaptive energy management features of the POWER7 chip, IEEE Micro 31 (2) (2011) 60–75.

[50] M. Floyd, M. Ware, K. Rajamani, T. Gloekler, B. Brock, P. Bose, A. Buyuktosunoglu, J.C. Rubio, B. Schubert, B. Spruth, J.A. Tierno, L. Pesantez, Adaptive energy-management features of the IBM POWER7 chip, IBM Journal of Research and Development 55 (3) (2011) 8:1–8:18.

[51] Vincent W. Freeh, David K. Lowenthal, Using multiple energy gears in MPI programs on a power-scalable cluster, in: Proceedings of the tenth ACM SIGPLAN symposium on Principles and practice of parallel programming, PPoPP '05, ACM, New York, NY, USA, 2005, pp. 164–173.

[52] V.W. Freeh, Feng Pan, N. Kappiah, D.K. Lowenthal, R. Springer, Exploring the energy-time tradeoff in MPI programs on a power-scalable cluster, in: Proceedings of the19th IEEE International on Parallel and Distributed Processing Symposium, 2005, April 2005, p. 4a.

[53] Wolfgang Frings, Morris Riedel, Llview: user-level monitoring in computational grids and e-science infrastructures, 2007.

[54] H. Fröning, M. Nüssle, H. Litz, U. Brüning, A case for FPGA based accelerated communication, in: Ninth International Conference on Networks (ICN) 2010, April 2010, pp. 28–33.

[55] Rong Ge, Xizhou Feng, Wu chun Feng, K.W. Cameron, Cpu miser: a performance-directed, run-time system for power-aware clusters, in: International Conference on Parallel Processing 2007, ICPP 2007, September 2007, pp. 18.

[56] Rong Ge, Xizhou Feng, Shuaiwen Song, Hung-Ching Chang, Dong Li, K.W. Cameron, Powerpack: energy profiling and analysis of high-performance systems and applications, IEEE Transactions on Parallel and Distributed Systems 21 (5) (2010) 658–671.

[57] Markus Geimer, Felix Wolf, Brian J.N. Wylie, Erika Ábrahám, Daniel Becker, Bernd Mohr, The Scalasca performance toolset architecture, Concurrency and Computation: Practice and Experience 22 (6) (2010) 702–719.

[58] P. Gepner, D.L. Fraser, M.F. Kowalik, R. Tylman, New multi-core intel xeon processors help design energy efficient solution for high performance computing, in: International Multiconference on Computer Science and Information Technology 2009, IMCSIT '09, October 2009, pp. 567–571.

[59] ZES ZIMMER Electronic Systems GmbH, 4-Channel Power Meter LMG450—Universal Meter for Motors, Power Electronics and Energy Analysis, February 2010.

[60] B. Goel, S.A. McKee, R. Gioiosa, K. Singh, M. Bhadauria, M. Cesati, Portable, scalable, per-core power estimation for intelligent resource management, in: International Green Computing Conference 2010, August 2010, pp. 135–146.

[61] G. Goldrian, T. Huth, B. Krill, J. Lauritsen, H. Schick, I. Ouda, S. Heybrock, D. Hierl, T. Maurer, N. Meyer, A. Schafer, S. Solbrig, T. Streuer, T. Wettig, D. Pleiter, K.-H. Sulanke, F. Winter, H. Simma, S.F. Schifano, R. Tripiccione, Qpace: quantum chromodynamics parallel computing on the cell broadband engine, Computing in Science Engineering 10 (6) (2008) 46–54.

[62] R. Gruber, V. Keller, E. Strohmaier, HPC@Green IT: Green High Performance Computing Methods, Springer, 2010.

[63] R.A. Haring, M. Ohmacht, T.W. Fox, M.K. Gschwind, D.L. Satterfield, K. Sugavanam, P.W. Coteus, P. Heidelberger, M.A. Blumrich, R.W. Wisniewski, A. Gara, G.L.-T. Chiu, P.A. Boyle, N.H. Chist, Changhoan Kim, The IBM blue Gene/Q compute chip, IEEE Micro 32 (2) (2012) 48–60.

[64] Michael Hennecke, Wolfgang Frings, Willi Homberg, Anke Zitz, Michael Knobloch, Hans Böttiger, Measuring power consumption on IBM blue Gene/P, Computer Science—Research and Development 1–8, doi: 10.1007/s00450-011-0192-y.

[65] S. Herbert, D. Marculescu, Analysis of dynamic voltage/frequency scaling in chip-multiprocessors, in: ACM/IEEE International Symposium on Low Power Electronics and Design (ISLPED) 2007, August 2007, pp. 38–43.

[66] Hewlett-Packard, HP power advisor utility: a tool for estimating power requirements for HP ProLiant server systems, 2009.

[67] Hewlett-Packard Corporation and Intel Corporation and Microsoft Corporation and Phoenix Technologies Ltd. and Toshiba Corporation, Advances Configuration and Power Interface Specification, Rev. 5.0, November 2011.

[68] L.P. Hewlett-Packard Development Company, Hp project moonshot: changing the game with extreme low-energy computing, 2012.

[69] T. Hoefler, Software and hardware techniques for power-efficient HPC networking, Computing in Science Engineering 12 (6) (2010) 30–37.

[70] T. Hoefler, M. Snir, Generic topology mapping strategies for large-scale parallel architectures, in: Proceedings of the 2011 ACM International Conference on Supercomputing (ICS'11), ACM, June 2011, pp. 75–85.

[71] Sunpyo Hong, Hyesoon Kim, An integrated GPU power and performance model, SIGARCH Computer Architecture News 38 (3) (2010) 280–289.

[72] Y. Hotta, M. Sato, H. Kimura, S. Matsuoka, T. Boku, D. Takahashi, Profile-based optimization of power performance by using dynamic voltage scaling on a pc cluster, in: 20th International Parallel and Distributed Processing Symposium 2006, IPDPS 2006, April 2006, p. 8.

[73] Chung hsing Hsu, Wu chun Feng, Jeremy S. Archuleta, Towards efficient supercomputing: a quest for the right metric, in: In Proceedings of the HighPerformance Power-Aware Computing Workshop, 2005.

[74] Chung-hsing Hsu, Wu-chun Feng, A power-aware run-time system for high-performance computing, in: Proceedings of the 2005 ACM/IEEE conference on Supercomputing, SC '05, IEEE Computer Society, Washington, DC, USA, 2005, p. 1.

[75] X.S. Hu, R.C. Murphy, S. Dosanjh, K. Olukotun, S. Poole, Hardware/software co-design for high performance computing: Challenges and opportunities, in: IEEE/ACM/IFIP International Conference on Hardware/Software Codesign and System Synthesis (CODES+ISSS) 2010, October 2010, pp. 63–64.

[76] Hai Huang, Kang G. Shin, Charles Lefurgy, Tom Keller, Improving energy efficiency by making dram less randomly accessed, in: Proceedings of the 2005 international symposium on Low power electronics and design, ISLPED '05, ACM, New York, NY, USA, 2005, pp. 393–398.

[77] S. Huang, S. Xiao, W. Feng, On the energy efficiency of graphics processing units for scientific computing, in: IEEE International Symposium on Parallel Distributed Processing 2009, IPDPS 2009, May 2009, pp. 1–8.

[78] R. Kalla, B. Sinharoy, W.J. Starke, M. Floyd, POWER7: IBM's next-generation server processor, IEEE Micro 30 (2) (2010) 7–15.

[79] K. Kandalla, E.P. Mancini, S. Sur, D.K. Panda, Designing power-aware collective communication algorithms for infiniband clusters, in: 39th International Conference on Parallel Processing (ICPP) 2010, September 2010, pp. 218–227.

[80] Mahmut Kandemir, Ibrahim Kolcu, Ismail Kadayif, Influence of loop optimizations on energy consumption of multi-bank memory systems, in: R. Horspool (Ed.), Compiler Construction, vol. 2304 of Lecture Notes in Computer Science, Springer Berlin/ Heidelberg, 2002, pp. 11–23, doi: 10.1007/3-540-45937-5.

[81] Mahmut Kandemir, Ugur Sezer, Victor Delaluz, Improving memory energy using access pattern classification, in: Proceedings of the 2001 IEEE/ACM international conference on Computer-aided design, ICCAD '01, IEEE Press, Piscataway, NJ, USA, 2001, pp. 201–206.

[82] J. Kavalieros, B. Doyle, S. Datta, G. Dewey, M. Doczy, B. Jin, D. Lionberger, M. Metz, W. Rachmady, M. Radosavljevic, U. Shah, N. Zelick, R. Chau, Tri-gate transistor architecture with high-k gate dielectrics, metal gates and strain engineering, in: Symposium on VLSI Technology 2006, Digest of Technical Papers 2006, 0-0 2006, pp. 50–51.

[83] H.S. Kim, M.J. Irwin, N. Vijaykrishnan, M. Kandemir, Effect of compiler optimizations on memory energy, in: IEEE Workshop on Signal Processing Systems 2000, SiPS 2000, 2000, pp. 663–672.

[84] J. Kim, K. Choi, G. Loh, Exploiting new interconnect technologies in on-chip communication, IEEE Journal on Emerging and Selected Topics in Circuits and Systems 2 (2) (2012) 124–136.

[85] H. Kimura, M. Sato, Y. Hotta, T. Boku, D. Takahashi, Emprical study on reducing energy of parallel programs using slack reclamation by DVFS in a power-scalable high performance cluster, in: IEEE International Conference on Cluster Computing 2006, September 2006, pp. 1–10.

[86] V.V. Kindratenko, J.J. Enos, Guochun Shi, M.T. Showerman, G.W. Arnold, J.E. Stone, J.C. Phillips, Wen mei Hwu, GPU clusters for high-performance computing, in: IEEE International Conference on Cluster Computing and Workshops 2009, CLUSTER '09, 31 2009-September 4 2009, pp. 1–8.

[87] Michael Knobloch, Timo Minartz, Daniel Molka, Stephan Krempel, Thomas Ludwig, Bernd Mohr, Electronic poster: eeclust: energy-efficient cluster computing, in: Proceedings of the 2011 companion on High Performance Computing Networking, Storage and Analysis Companion SC '11 Companion, ACM, New York, NY, USA, 2011, pp. 99–100.

[88] Michael Knobloch, Bernd Mohr, Timo Minartz, Determine energy-saving potential in wait-states of large-scale parallel programs, Computer Science—Research and Development (2011).

[89] Andreas Knüpfer, Holger Brunst, Jens Doleschal, Matthias Jurenz, Matthias Lieber, Holger Mickler, Matthias S. Müller, Wolfgang E. Nagel, The vampir performance analysis tool-set, in: Proceedings of the 2nd International Workshop on Parallel Tools, Tools for High Performance Computing, Springer, 2008, pp. 139–155.

[90] A. Kodi, A. Louri, Reconfigurable and adaptive photonic networks for high-performance computing systems, Applied Optics 48 (2009) 13.

[91] A.K. Kodi, A. Louri, Energy-efficient and bandwidth-reconfigurable photonic networks for high-performance computing (HPC) systems, IEEE Journal of Selected Topics in Quantum Electronics 17 (2) (2011) 384–395.

[92] Julian Kunkel, Olga Mordvinova, Michael Kuhn, Thomas Ludwig, Collecting energy consumption of scientific data, Computer Science—Research and Development 25 (2010) 197–205, doi: 10.1007/s00450-010-0121-5.

[93] N.A. Kurd, S. Bhamidipati, C. Mozak, J.L. Miller, P. Mosalikanti, T.M. Wilson, A.M. El-Husseini, M. Neidengard, R.E. Aly, M. Nemani, M. Chowdhury, R. Kumar, A family of 32 nm IA processors, IEEE Journal of Solid-State Circuits 46 (1) (2011) 119–130.

[94] J.S. Kuskin, C. Young, J.P. Grossman, B. Batson, M.M. Deneroff, R.O. Dror, D.E. Shaw, Incorporating flexibility in anton, a specialized machine for molecular dynamics simulation, in: IEEE 14th International Symposium on High Performance Computer Architecture 2008, HPCA 2008, Febraury 2008, pp. 343–354.

[95] K.-D. Lange, Identifying shades of green: the specpower benchmarks, Computer 42 (3) (2009) 95–97.

[96] R.H. Larson, J.K. Salmon, R.O. Dror, M.M. Deneroff, C. Young, J.P. Grossman, Yibing Shan, J.L. Klepeis, D.E. Shaw, High-throughput pairwise point interactions in anton, a specialized machine for molecular dynamics simulation, in: IEEE 14th International Symposium on High Performance Computer Architecture 2008, HPCA 2008, Febraury 2008, pp. 331–342.

[97] Jungseob Lee, V. Sathisha, M. Schulte, K. Compton, Nam Sung Kim, Improving throughput of power-constrained GPUS using dynamic voltage/frequency and core scaling, in: International Conference on Parallel Architectures and Compilation Techniques (PACT) 2011, October 2011, pp. 111–120.

[98] J. Leverich, M. Monchiero, V. Talwar, P. Ranganathan, C. Kozyrakis, Power management of datacenter workloads using per-core power gating, Computer Architecture Letters 8 (2) (2009) 48–51.

[99] Dong Li, B.R. de Supinski, M. Schulz, K. Cameron, D.S. Nikolopoulos, Hybrid MPI/OpenMP power-aware computing, in: IEEE International Symposium on Parallel Distributed Processing (IPDPS) 2010, April 2010, pp. 1–12.

[100] Dong Li, D.S. Nikolopoulos, K. Cameron, B.R. de Supinski, M. Schulz, Power-aware MPI task aggregation prediction for high-end computing systems, in: IEEE International Symposium on Parallel Distributed Processing (IPDPS) 2010, April 2010, pp. 1–12.

[101] Min Yeol Lim, Vincent W. Freeh, David K. Lowenthal, Adaptive, transparent frequency and voltage scaling of communication phases in MPI programs, in: SC 2006 Conference, Proceedings of the ACM/IEEE, November 2006, p. 14.

[102] M.Y. Lim, V.W. Freeh, Determining the minimum energy consumption using dynamic voltage and frequency scaling, in: IEEE International Parallel and Distributed Processing Symposium 2007, IPDPS 2007, March 2007, pp. 1–8.

[103] H. Litz, H. Fröning, M. Nüssle, U. Brüning, Velo: a novel communication engine for ultra-low latency message transfers, in: 37th International Conference on Parallel Processing 2008, ICPP '08, September 2008, pp. 238–245.

[104] Jiuxing Liu, Dan Poff, Bulent Abali, Evaluating high performance communication: a power perspective, in: Proceedings of the 23rd international conference on Supercomputing, ICS '09, ACM, New York, NY, USA, 2009, pp. 326–337.

[105] Owen Liu, Amd technology: power, performance and the future, in: Proceedings of the 2007 Asian technology information program's (ATIP's) 3rd workshop on High performance computing in China: solution approaches to impediments for high performance computing, CHINA HPC '07, ACM, New York, NY, USA, 2007, pp. 89–94.

[106] Charles Lively, Xingfu Wu, Valerie Taylor, Shirley Moore, Hung-Ching Chang, Chun-Yi Su, Kirk Cameron, Power-aware predictive models of hybrid (MPI/OPENMP) scientific applications on multicore systems, Computer Science—Research and Development 1–9, doi: 10.1007/s00450-011-0190-0.

[107] Hatem Ltaief, Piotr Luszczek, Jack Dongarra, Profiling high performance dense linear algebra algorithms on multicore architectures for power and energy efficiency, Computer Science—Research and Development 1–11, doi: 10.1007/s00450-011-0191-z.

[108] Anita Lungu, Pradip Bose, Alper Buyuktosunoglu, Daniel J. Sorin, Dynamic power gating with quality guarantees, in: Proceedings of the 14th ACM/IEEE international symposium on Low power electronics and design, ISLPED '09, ACM, New York, NY, USA, 2009, pp. 377–382.

[109] Piotr R Luszczek, David H Bailey, Jack J Dongarra, Jeremy Kepner, Robert F Lucas, Rolf Rabenseifner, Daisuke Takahashi, The HPC challenge (HPCC) benchmark suite, in: Proceedings of the 2006 ACM/IEEE conference on Supercomputing, SC '06, ACM, New York, NY, USA, 2006.

[110] K. Malkowski, Ingyu Lee, P. Raghavan, M.J. Irwin, On improving performance and energy profiles of sparse scientific applications, in: 20th International Parallel and Distributed Processing Symposium 2006, IPDPS 2006, April 2006, pp. 8.

[111] Satoshi Matsuoka, Takayuki Aoki, Toshio Endo, Akira Nukada, Toshihiro Kato, Atushi Hasegawa, GPU accelerated computing–from hype to mainstream, the rebirth of vector computing, Journal of Physics: Conference Series 180 (1) (2009).

[112] H.-Y. McCreary, M.A. Broyles, M.S. Floyd, A.J. Geissler, S.P. Hartman, F.L. Rawson, T.J. Rosedahl, J.C. Rubio, M.S. Ware, Energyscale for IBM POWER6 microprocessor-based systems, IBM Journal of Research and Development 51 (6) (2007) 775–786.

[113] Huzefa Mehta, Robert Michael Owens, Mary Jane Irwin, Rita Chen, Debashree Ghosh, Techniques for low energy software, in: Proceedings of the 1997 international symposium on Low power electronics and design, ISLPED '97, ACM, New York, NY, USA, 1997, pp. 72–75.

[114] Timo Minartz, Michael Knobloch, Thomas Ludwig, Bernd Mohr, Managing hardware power saving modes for high performance computing, in: International Green Computing Conference and Workshops (IGCC) 2011, 2011, pp. 1–8.

[115] Timo Minartz, Julian Kunkel, Thomas Ludwig, Simulation of power consumption of energy efficient cluster hardware, Computer Science—Research and Development, 25 (2010) 165–175, doi: 10.1007/s00450-010-0120-6.

[116] Timo Minartz, Daniel Molka, Michael Knobloch, Stephan Krempel, Thomas Ludwig, Wolfgang E. Nagel, Bernd Mohr, Hugo Falter, eeclust: energy-efficient cluster computing, in: Christian Bischof, Heinz-Gerd Hegering, Wolfgang E. Nagel, Gabriel Wittum (Eds.), Competence in High Performance Computing 2010, Springer Berlin Heidelberg, 2012, pp. 111–124, doi: 10.1007/978-3-642-24025-6_10.

[117] Lauri Minas, Brad Ellison, Power metrics for data centers, in: Energy Efficiency for Information Technology—How to Reduce Power Consumption in Servers and Data Centers, Intel Press, 2009.

[118] C. Minkenberg, F. Abel, P. Muller, Raj Krishnamurthy, M. Gusat, P. Dill, I. Iliadis, R. Luijten, R.R. Hemenway, R. Grzybowski, E. Schiattarella, Designing a crossbar scheduler for hpc applications, IEEE Micro 26 (3) (2006) 58–71.

[119] D. Molka, D. Hackenberg, R. Schone, M.S. Muller, Characterizing the energy consumption of data transfers and arithmetic operations on x86-64 processors, in: International Green Computing Conference 2010, August 2010, pp. 123–133.

[120] Daniel Molka, Daniel Hackenberg, Robert Schöne, Timo Minartz, Wolfgang E. Nagel, Flexible Workload Generation for HPC Cluster Efficiency Benchmarking. Computer Science—Research and Development (2011).

[121] Gordon E. Moore, Cramming more components onto integrated circuits, Electronics 38 (8) (1965).

[122] MPI Forum. MPI: A Message-Passing Interface Standard. Version 2.2, September 4th 2009, December 2009, available at: <http://www.mpi-forum.org>.

[123] Trevor Mudge, Power: a first class design constraint for future architectures, in: Mateo Valero, Viktor Prasanna, Sriram Vajapeyam (Eds.), High Performance Computing—HiPC 2000, volume 1970 of Lecture Notes in Computer Science, Springer Berlin/Heidelberg, 2000, pp. 215–224, doi: 10.1007/3-540-44467-X_20.

[124] Matthias S. Müller, Andreas Knüpfer, Matthias Jurenz, Matthias Lieber, Holger Brunst, Hartmut Mix, Wolfgang E. Nagel, Developing scalable applications with vampir, vampirserver and vampirtrace, in: Parallel Computing: Architectures, Algorithms and Applications, vol. 15 of Advances in Parallel Computing, IOS Press, 2007, pp. 637–644.

[125] Olli Mämmelä, Mikko Majanen, Robert Basmadjian, Hermann De Meer, André Giesler, Willi Homberg, Energy-aware job scheduler for high-performance computing, Computer Science—Research and Development 1–11, doi: 10.1007/s00450-011-0189-6.

[126] Hitoshi Nagasaka, Naoya Maruyama, Akira Nukada, Toshio Endo, Satoshi Matsuoka, Statistical power modeling of GPU kernels using performance counters, in: Proceedings of the International Conference on Green Computing, GREENCOMP '10, IEEE Computer Society, Washington, DC, USA, 2010, pp. 115–122.

[127] M. Nüssle, B. Geib, H. Fröning, U. Brüning, An FPGA-based custom high performance interconnection network, in: International Conference on Reconfigurable Computing and FPGAs 2009, ReConFig '09, December 2009, pp. 113–118.

[128] Stuart S.P. Parkin, Masamitsu Hayashi, Luc Thomas, Magnetic domain-wall racetrack memory, Science 320 (5873) (2008) 190–194.

[129] J.T. Pawlowski. Hybrid memory cube: breakthrough dram perfonnance with a fundamentally re-architected dram subsystem, in: Hot Chips, 2011.

[130] F. Poletti, A. Poggiali, D. Bertozzi, L. Benini, P. Marchal, M. Loghi, M. Poncino, Energy-efficient multiprocessor systems-on-chip for embedded computing: exploring programming models and their architectural support, IEEE Transactions on Computers 56 (5) (2007) 606–621.

[131] Moinuddin K. Qureshi, Vijayalakshmi Srinivasan, Jude A. Rivers, Scalable high performance main memory system using phase-change memory technology, in: Proceedings of the 36th annual international symposium on Computer architecture, ISCA '09, ACM, New York, NY, USA, 2009, pp. 24–33.

[132] Alan Ross, Bob Stoddard, Data Center Energy Efficiency with Intel Power Management Technologies, 2010.

[133] E. Rotem, A. Naveh, D. Rajwan, A. Ananthakrishnan, E. Weissmann, Power-management architecture of the intel microarchitecture code-named sandy bridge, IEEE Micro 32 (2) (2012) 20–27.

[134] Barry Rountree, David K. Lowenthal, Shelby Funk, Vincent W. Freeh, Bronis R. de Supinski, Martin Schulz, Bounding energy consumption in large-scale MPI programs, in: Proceedings of the 2007 ACM/IEEE conference on Supercomputing, SC '07, ACM, New York, NY, USA, 2007, pp. 49:1–49:9.

[135] Barry Rountree, David K. Lownenthal, Bronis R. de Supinski, Martin Schulz, Vincent W. Freeh, Tyler Bletsch, Adagio: making dvs practical for complex HPC applications, in: Proceedings of the 23rd international conference on Supercomputing, ICS '09, ACM, New York, NY, USA, 2009, pp. 460–469.

[136] Richard M. Russell, The cray-1 computer system, ACM Communication 21 (1) (1978) 63–72.

[137] S. Rusu, S. Tam, H. Muljono, J. Stinson, D. Ayers, J. Chang, R. Varada, M. Ratta, S. Kottapalli, S. Vora, A 45 nm 8-core enterprise xeon processor, IEEE Journal of Solid-State Circuits 45 (1) (2010) 7–14.

[138] D. Schinke, N. Di Spigna, M. Shiveshwarkar, P. Franzon, Computing with novel floating-gate devices, Computer 44 (2) (2011) 29–36.

[139] Robert Schöne, Daniel Hackenberg, Daniel Molka, Simultaneous multithreading on x86_64 systems: an energy efficiency evaluation, in: Proceedings of the 4th Workshop on Power-Aware Computing and Systems, HotPower '11, ACM, New York, NY, USA, 2011, pp. 10:1–10:5.

[140] Robert Schöne, Ronny Tschüter, Thomas Ilsche, Daniel Hackenberg, The vampirtrace plugin counter interface: introduction and examples, in: Proceedings of the

2010 conference on Parallel processing, Euro-Par 2010, Berlin, Heidelberg, 2011, Springer-Verlag, pp. 501–511.

[141] T. Scogland, H. Lin, W. Feng, A first look at integrated GPUs for green high-performance computing. Computer Science—Research and Development 25 (2010) 125–134, doi: 10.1007/s00450-010-0128-y.

[142] Tom Scogland, Balaji Subramaniam, Wu-chun Feng, The green500 list: escapades to exascale, Computer Science—Research and Development 1–9, doi: 10.1007/s00450-012-0212-6.

[143] Seagate, Barracuda 7200.12 Serial ATA, February 2009.

[144] S. Sharma, Chung-Hsing Hsu, Wu chun Feng, Making a case for a green500 list, in: 20th International Parallel and Distributed Processing Symposium 2006, IPDPS 2006, April 2006, p. 8.

[145] David E. Shaw, Martin M. Deneroff, Ron O. Dror, Jeffrey S. Kuskin, Richard H. Larson, John K. Salmon, Cliff Young, Brannon Batson, Kevin J. Bowers, Jack C. Chao, Michael P. Eastwood, Joseph Gagliardo, J.P. Grossman, C. Richard Ho, Douglas J. Ierardi, István Kolossváry, John L. Klepeis, Timothy Layman, Christine McLeavey, Mark A. Moraes, Rolf Mueller, Edward C. Priest, Yibing Shan, Jochen Spengler, Michael Theobald, Brian Towles, Stanley C. Wang, Anton, a special-purpose machine for molecular dynamics simulation, in: Proceedings of the 34th annual international symposium on Computer architecture, ISCA '07, ACM, New York, NY, USA, 2007, pp. 1–12.

[146] Karan Singh, Major Bhadauria, Sally A. McKee, Real time power estimation and thread scheduling via performance counters, SIGARCH Computer Architecture News 37 (2) (2009) 46–55.

[147] Robert Springer, David K. Lowenthal, Barry Rountree, Vincent W. Freeh, Minimizing execution time in MPI programs on an energy-constrained, power-scalable cluster, in: Proceedings of the eleventh ACM SIGPLAN symposium on Principles and practice of parallel programming, PPoPP '06, ACM, New York, NY, USA, 2006, pp. 230–238.

[148] Bob Steigerwald, Christopher D. Lucero, Chakravarthy Akella, Abhishek Agrawal, Impact of software on energy consumption, in: Energy Aware Computing—Powerful Approaches for Green System Design, Intel Press, 2012.

[149] Bob Steigerwald, Christopher D. Lucero, Chakravarthy Akella, Abhishek Agrawal, Writing energy-efficient software, in: Energy Aware Computing—Powerful Approaches for Green System Design, Intel Press, 2012.

[150] Thomas Sterling, Donald J. Becker, Daniel Savarese, John E. Dorband, Udaya A. Ranawake, Charles V. Packer, Beowulf: a parallel workstation for scientific computation, in: In Proceedings of the 24th International Conference on Parallel Processing, CRC Press, 1995, pp. 11–14.

[151] Dmitri B. Strukov, Gregory S. Snider, Duncan R. Stewart, R. Stanley Williams, The missing memristor found, Nature 453 (7191) (2008) 80–83 (10.1038/nature06932).

[152] V. Sundriyal, M. Sosonkina, Fang Liu, M.W. Schmidt, Dynamic frequency scaling and energy saving in quantum chemistry applications, in: IEEE International Symposium on Parallel and Distributed Processing Workshops and Phd Forum (IPDPSW), 2011, May 2011, pp. 837–845.

[153] Vaibhav Sundriyal, Masha Sosonkina. Per-call energy saving strategies in all-to-all communications, in: Yiannis Cotronis, Anthony Danalis, Dimitrios Nikolopoulos, Jack Dongarra (Eds.), Recent Advances in the Message Passing Interface, vol. 6960 of Lecture Notes in Computer Science, Springer Berlin/Heidelberg, 2011, pp. 188–197, doi: 10.1007/978-3-642-24449-0_22.

[154] Dan Terpstra, Heike Jagode, Haihang You, Jack Dongarra, Collecting performance data with papi-c, in: Matthias S. Müller, Michael M. Resch, Alexander Schulz, Wolfgang E. Nagel (Eds.), Tools for High Performance Computing 2009, Springer Berlin Heidelberg, 2010, pp. 157–173, doi: 10.1007/978-3-642-11261-4_11.

[155] Niraj Tolia, Zhikui Wang, Manish Marwah, Cullen Bash, Parthasarathy Ranganathan, Xiaoyun Zhu, Delivering energy proportionality with non energy-proportional systems: optimizing the ensemble, in: Proceedings of the 2008 conference on Power aware computing and systems, HotPower'08, USENIX Association, Berkeley, CA, USA, 2008, pp. 2–2.

[156] J. Torrellas, Architectures for extreme-scale computing, Computer 42 (11) (2009) 28–35.

[157] Abhinav Vishnu, Shuaiwen Song, Andres Marquez, Kevin Barker, Darren Kerbyson, Kirk Cameron, Pavan Balaji, Designing energy efficient communication runtime systems: a view from PGAS models, The Journal of Supercomputing pp. 1–19, doi: 10.1007/s11227-011-0699-9.

[158] Malcolm Ware, Karthick Rajamani, Michael Floyd, Bishop Brock, Juan C Rubio, Freeman Rawson, John B. Carter, Architecting for power management: the IBM® POWER7$^{TM}$ approach, in: The Sixteenth International Symposium on HighPerformance Computer Architecture, HPCA 16, 2010, pp. 1–11.

[159] M.S. Warren, E.H. Weigle, Wu-Chun Feng, High-density computing: A 240-processor beowulf in one cubic meter, in: ACM/IEEE 2002 Conference on Supercomputing, November 2002, p. 61.

[160] Michael Wehner, Leonid Oliker, John Shalf, Towards ultra-high resolution models of climate and weather, International Journal of High Performance Computing Applications 22 (2) (2008) 149–165.

[161] Mathias Winkel, Robert Speck, Helge Hübner, Lukas Arnold, Rolf Krause, Paul Gibbon, A massively parallel, multi-disciplinary barnes–hut tree code for extreme-scale n-body simulations, Computer Physics Communications 183 (4) (2012) 880–889.

[162] Felix Wolf, Brian J.N. Wylie, Erika Ábrahám, Daniel Becker, Wolfgang Frings, Karl Fürlinger, Markus Geimer, Marc-André Hermanns, Bernd Mohr, Shirley Moore, Matthias Pfeifer, Zoltán Szebenyi, Usage of the scalasca toolset for scalable performance analysis of large-scale parallel applications, in: Michael Resch, Rainer Keller, Valentin Himmler, Bettina Krammer, Alexander Schulz (Eds.), Tools for High Performance Computing, Springer Berlin Heidelberg, 2008, pp. 157–167, doi: 10.1007/978-3-540-68564-7_10.

[163] H-S Philip Wong, Simone Raoux, SangBum Kim, Jiale Liang, John P. Reifenberg, Bipin Rajendran, Mehdi Asheghi, Kenneth E Goodson, Phase change memory, in: Proceedings of the IEEE, vol. 98, 2010, pp. 2201–2227.

[164] Katherine Yelick, Dan Bonachea, Wei-Yu Chen, Phillip Colella, Kaushik Datta, Jason Duell, Susan L. Graham, Paul Hargrove, Paul Hilfinger, Parry Husbands, Costin Iancu, Amir Kamil, Rajesh Nishtala, Jimmy Su, Michael Welcome, Tong Wen, Productivity and performance using partitioned global address space languages, in: Proceedings of the 2007 international workshop on Parallel symbolic computation, PASCO '07, ACM, New York, NY, USA, 2007, pp. 24–32.

[165] R. Zamani, A. Afsahi, Ying Qian, C. Hamacher, A feasibility analysis of power-awareness and energy minimization in modern interconnects for high-performance computing, in: IEEE International Conference on Cluster Computing, 2007, September 2007, pp. 118–128.

[166] Severin Zimmermann, Ingmar Meijer, Manish K. Tiwari, Stephan Paredes, Bruno Michel, Dimos Poulikakos, Aquasar: a hot water cooled data center with direct energy reuse, Energy 43 (1) (2012) 237–245 (2nd International Meeting on Cleaner Combustion (CM0901-Detailed Chemical Models for Cleaner Combustion)).

## ABOUT THE AUTHOR

**Michael Knobloch** received his diploma in Mathematics from Technische Universität Dresden, Germany in 2008. Since 2009 he holds a position as researcher at the Jülich Supercompuing Centre (JSC) of Forschungszentrum Jülich GmbH in the performance analysis group led by Dr. Bernd Mohr. His research interests include energy and performance analysis of HPC applications and the development of corresponding tools. He is part of several Exascale efforts at JSC.

**CHAPTER TWO**

# Micro-Fluidic Cooling for Stacked 3D-ICs: Fundamentals, Modeling and Design

**Bing Shi  and  Ankur Srivastava**
University of Maryland, College Park, MD, USA

## Contents

*Advances in Computers*, Volume 88
ISSN 0065-2458, http://dx.doi.org/10.1016/B978-0-12-407725-6.00002-2

## Abstract

The three-dimensional integrated circuit (3D-IC), which enables better integration density, faster on-chip communications and heterogenous integration, etc., has become
an active topic of research. Despite its significant performance improvement over the
conventional 2D circuits, 3D-IC also exhibits thermal issues due to its high power density caused by the stacked architecture. To fully exploit the benefit of 3D-ICs, future
3D-IC designs are expected to have significantly complex architectures and integration
levels that would be associated with very high power dissipation and heat density.
The conventional air cooling has already been proved insufficient for cooling stacked
3D-ICs since several microprocessors are stacked vertically, and the interlayer micro-
channel liquid cooling provides a better option to address this problem. This chapter
investigates several aspects of 3D-ICs with micro-channel heat sinks including their
infrastructure, thermal and hydrodynamic modeling, current research achievements,
design challenges, etc. We will also introduce a micro-channel based runtime thermal
management approach which dynamically controls the 3D-IC temperature by controlling the fluid flow rate through micro-channels.

## Nomenclature

| | |
|---|---|
| $A$ | surface area of heat transfer in micro–channel |
| $C_v$ | volumetric specific heat of coolant |
| $D_h$ | hydraulic diameter of micro–channel |
| $F$ | volumetric fluid flow rate |
| $G$ | fluid mass flux |
| $H$ | thermal conductance matrix |
| $K_d$ | pressure loss coefficient for fluid developing region |
| $K_{90°}$ | pressure loss coefficient for 90 °C bend |
| $L$ | length of micro–channel |
| $L_d$ | length of fluid developing region |
| $L_f$ | total length of fully developed region in a bended micro–channel |
| $P$ | fluid pressure |

| | |
|---|---|
| $\Delta P$ | pressure drop across micro-channels |
| $\Delta P_f$ | total pressure drop of all fully developed region in a bended micro-channel |
| $\Delta P_d$ | total pressure drop of all fluid developing region in a bended micro-channel |
| $\Delta P_{90°}$ | total pressure drop of 90°C bend in a bended micro-channel |
| $Q$ | power profile of a 3D-IC |
| $Q_{pump}$ | micro-channel pumping power |
| $R_{conv}$ | convective resistance |
| $R_{heat}$ | heat resistance |
| $R_{cond}$ | conductive resistance |
| $S$ | surface of a fluid control volume |
| $T$ | temperature (thermal profile) |
| $T_{max}$ | maximum thermal constraint |
| $T_{f,j}/T_{w,j}$ | temperature of fluid/solid layer in $j$th tier |
| $T_{f,i}/T_{f,o}$ | fluid temperature at micro-channel inlet/outlet |
| $T_{w,i}/T_{w,o}$ | temperature of micro-channel wall at inlet/outlet plane |
| $T_{i,j,k}$ | temperature at $i$th/$j$th/$k$th grid in $x/y/z$ direction |
| $T_{s1}/T_{s2}$ | surface temperature at the front and back of a grid |
| $U$ | control volume of fluid |
| $V$ | fluid velocity |
| $b$ | micro-channel width |
| $d$ | micro-channel height |
| $e$ | fluid enthalpy |
| $f_r$ | fraction factor |
| $h$ | convective heat transfer coefficient |
| $k_f$ | thermal conductivity of fluid |
| $k_{f,x/y/z}$ | fluid thermal conductivity in $x/y/z$ direction |
| $k_w$ | thermal conductivity of solid layer |
| $l$ | characteristic length |
| $\Delta p_d$ | pressure drop of a fluid developing section |
| $\Delta p_{90°}$ | pressure drop at each 90 °C bend corner |
| $q$ | power dissipation |
| $s_{rr,ij}$ | radial stress applied on gate $j$ caused by $i$th TSV |
| $t$ | time |
| $u$ | vapor quality |
| $\alpha$ | carrier mobility |
| $\beta$ | micro-channel aspect ratio |
| $\rho$ | fluid density |

| $\mu$ | fluid viscosity |
| $\eta_0$ | surface coefficient |
| $\omega_{ij}$ | angle between the line connecting the center of TSV $i$ and transistor $j$ and the channel direction of transistor $j$ |
| $\psi$ | orientation coefficient indicating the impact of $\omega$ on carrier mobility |

## 1. INTRODUCTION

The three-dimensional integrated circuit (3D-IC), which consists of several vertically stacked layers of active electronic components, provides faster on-chip communications compared with equivalent 2D circuits. They also result in overall system energy savings, increased integration densities, and co-integration of heterogenous components. Despite its significant performance improvement over 2D circuits, 3D-IC also exhibits thermal issues due to its high power density caused by the stacked architecture. Although current 3D-IC designs are limited to partitioning of memory and datapath across layers, future 3D-IC designs are expected to have significantly complex architectures and integration levels that would be associated with very high power dissipation and heat density.

The conventional air cooling might not be enough for cooling stacked 3D-ICs since several microprocessors are stacked vertically. As illustrated in [1], if two $100 \, \text{W}/\text{cm}^2$ microprocessors are stacked on top of each other, the power density becomes $200 \, \text{W}/\text{cm}^2$, which is beyond the heat removal limits of air cooled heat sinks. However, the interlayer micro-channel liquid cooling provides a better option to address this problem. Micro-channel cooling integrates micro-channel heat sinks into each tier of the 3D-IC and uses liquid flow to remove heat inside 3D chip. It has great capability of cooling high heat density (as much as $700 \, \text{W}/\text{cm}^2$ [2]) and therefore is very appropriate for cooling 3D-ICs.

This chapter investigates several aspects of 3D-ICs with micro-channel heat sinks including their infrastructure, thermal modeling, design challenges, etc. In Section 2, we introduce the fundamental characteristics of fluid in micro-channels. We then discuss the challenges for designing the micro-channel heat sink in 3D-ICs in Section 3. We summarize the existing thermal and hydrodynamic modeling approaches for 3D-IC with micro-channels, as well as the existing works on micro-channel design and optimization in Sections 4 and 5. Section 6 introduces a micro-channel based runtime thermal management approach which dynamically controls the 3D-IC temperature by controlling the fluid flow rate through micro-channels.

## 2. FUNDAMENTAL CHARACTERISTICS OF FLUIDS IN MICRO-CHANNELS

### 2.1 3D-IC Structure with Micro-Channel Cooling

Figure 1 shows a three-tier 3D-IC with interlayer micro-channel heat sinks. Each tier contains an active silicon layer which consists of functional units such as cores and memories. These electronic components dissipate power. Micro-channel heat sinks are embedded in the bulk silicon layer below each active silicon layer to provide cooling [3, 4].

As shown in Fig. 1, the fluid pump pushes the cold coolant fluid into the micro-channels. The coolant then goes through the micro-channels (in $z$ direction), taking away the heat inside the 3D-IC. The heated coolant is then cooled down in the heat exchanger, and recirculates into the fluid pump again for the cooling in the next circulation.

Heat removal through micro-channels comprises of a combination of conduction, convection, and coolant flow. Heat dissipated in surrounding regions conducts to the micro-channel sidewalls which is then absorbed by

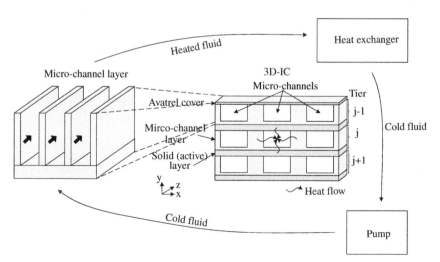

**Fig. 1.** Stacked 3D-IC with micro-channel cooling system.

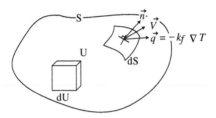

**Fig. 2.** Control volume of fluid.

the flowing fluid through convection. The heated fluid is then carried away by the moving flow [5].

## 2.2 Conservation Law of Fluid Dynamics

The characteristic of fluid inside the micro-channels is governed by conservation law of fluid. Considering the control volume of fluid $U$ and its surface $S$ (as shown in Fig. 2). The fluid flow in the control volume is governed by the following mass, momentum, and energy conservation equations [6–10]:

$$\text{Mass conservation:} \quad \frac{\partial \rho}{\partial t} + \nabla \cdot (\rho \vec{V}) = 0,$$

$$\text{Momentum conservation:} \quad \rho \left( \frac{\partial \vec{V}}{\partial t} + \vec{V} \cdot \nabla \vec{V} \right) = -\nabla P + \mu \nabla^2 \vec{V},$$

$$\text{Energy conservation:} \quad C_v \frac{dT}{dt} + \nabla \cdot (- k_f \nabla T) + C_v \vec{V} \cdot \nabla T = \dot{q}. \tag{1}$$

Here $\vec{V}$ is the flow velocity vector, $T$ is the fluid temperature, $\dot{q}$ is the volumetric heat generation rate, and $P$ is the pressure inside fluid. Also, $\rho, \mu, C_v$, and $k_f$ are the density, viscosity, volumetric specific heat, and thermal conductivity of the fluid, respectively.

## 2.3 Dimensionless Numbers in Fluid Mechanics

The governing equations above are complex partial differential equations (PDE). Researchers in fluid mechanics introduced a set of dimensionless numbers which could help simplify the complex problem also better understand the relative importance of forces, energies, or time scales [6, 11]. Some of these dimensionless numbers are Reynolds number ($Re$), Prandtl number ($Pr$), Nusselt number ($Nu$), etc.

*Reynolds number Re:* The Reynolds number gives a measure of the ratio between inertial forces to viscous forces, and is defined as:

$$Re = \frac{\rho V l}{\mu}, \tag{2}$$

where $V$ is the mean fluid velocity and $l$ is the characteristic length. In straight micro-channels, the characteristic length is usually given by the hydraulic diameter $D_h$. When the cross section of the channel is circular, $D_h$ is the diameter of the cross section, while in rectangular channels, $D_h$ is defined as $D_h = 4 \cdot$ cross sectional area/perimeter $= 4bd/(2b + 2d)$, where $b$ and $d$ are the width and height of the micro-channel. Usually, the Reynolds number is used to distinguish between laminar and turbulent flow, which will be explained later.

*Prandtl number Pr:* The Prandtl number is the ratio of momentum diffusivity (kinematic viscosity) to thermal diffusivity:

$$Pr = \frac{\text{kinematic viscosity}}{\text{thermal diffusivity}} = \frac{\mu/\rho}{k_f/(\rho C_v)} = \frac{C_v \mu}{k_f}. \tag{3}$$

*Nusselt number Nu:* The Nusselt number is the ratio of convective to conductive heat transfer across the boundary between the fluid and solid. The Nusselt number is defined as:

$$Nu = \frac{hl}{k_f}, \tag{4}$$

where $h, l,$ and $k_f$ are the convective heat transfer coefficient, channel characteristic length, and fluid thermal conductivity. Usually, $N_u$ is used to calculate the convective heat transfer coefficient $h$. Many works have been done to characterize the Nusselt number in micro-channels, and express it as a function of the Reynolds number and Prandtl number [12–15].

## 2.4 3D-IC with Micro-Channels: Modeling

Based on the conservation law in Eqn (1), the works in [3, 4] simplify the energy and momentum conservation formulations of micro-channels in 3D-ICs (assuming the working fluid is incompressible steady flow). To simplify the formulation, they discretize the 3D-IC into tiers in $y$ direction and each tier is partitioned into solid layer and micro-channel layer (as Fig. 1 shows). Also, each micro-channel layer is discretized into fluid and solid part in $x$ direction.

### 2.4.1 Energy Conservation

Let $T_{w,j}(x,z)$ represent the temperature of a point in the solid layer of the $j$th tier, and the location of this point in $x$ and $z$ direction is $(x,z)$. Similarly, $T_{f,j}(x,z)$ represents the temperature of a point in the fluid of micro-channel layer of the $j$th tier, and the location of this point in $x$ and $z$ direction is $(x,z)$. The energy conservation can be described as:

Micro-channel layer:

$$\rho F \frac{de}{dz} = \eta_0 h(2b + 2d)(T_{w,j}(x,z) - T_{f,j}(x,z))$$
$$+ hb(T_{w,j+1}(x,z) - T_{f,j}(x,z)).$$

Solid layer:

$$-\left(\frac{\partial}{\partial x}\left(k_w \frac{\partial T_{w,j}(x,z)}{\partial x}\right) + \frac{\partial}{\partial y}\left(k_w \frac{\partial T_{w,j}(x,z)}{\partial y}\right)\right.$$
$$\left. + \frac{\partial}{\partial z}\left(k_w \frac{\partial T_{w,j}(x,z)}{\partial z}\right)\right) = \dot{q}_a + \dot{q}_{conv}. \tag{5}$$

In Eqn (5), the first equation is the energy conservation law in micro-channel fluid. The left-hand side measures the fluid enthalpy change due to the heat convection from the solid, where $e$ is the fluid enthalpy per unit mass, and $F$ is the volumetric fluid flow rate. The volumetric flow rate in a micro-channel $F =$ velocity $*$ cross sectional area $= Vbd$ ($V$ is the velocity of fluid in micro-channel, $b$ and $d$ are the width and height of the channel). The right-hand side measures the heat transferred from the solid layer (both above and below) to the fluid layer. Here $\eta_0$ is the surface efficiency of the micro-channel sidewalls and bases, and $(2b + 2d)$ in the first term on the right-hand side represents the heat transfer area between fluid and solid layer in the same tier, while $b$ in the second term represents the heat transfer area between fluid and solid layer in different but neighboring tiers.

The second equation in Eqn (5) is the energy conservation in solid layer. The left-hand side represents the heat loss due to heat conduction ($k_w$ is the thermal conductivity of the solid). This heat conduction has two source/sink terms which are given on the right-hand side of this equation, where $\dot{q}_a$ is the heat generated in the active layer, and $\dot{q}_{conv}$ is the heat convected to the fluid.

### 2.4.2 Momentum Conservation

The pressure drop in the fluid $\Delta P$ is caused by frictional forces on the fluid as it flows through the micro-channel. Pressure drop along the micro-channel in the fluid direction $z$ is:

$$-\frac{dP}{dz} = \frac{2f_r G^2}{\rho D_h} + \frac{d}{dz}\left(\frac{G^2}{\rho}\right). \tag{6}$$

Here, $f_r$ is the friction factor whose value depends on the flow conditions, the micro-channel geometry, and surface conditions, and $G$ is the flow mass flux.

The fluid density $\rho$ depends on whether the fluid is single phase or two phase (which will be explained in the following subsection). For single phase, the fluid density is just the density of the working liquid. For two phase flow, since the working fluid is a mixture of both liquid and vapor, the density is:

$$\frac{1}{\rho} = \frac{1-u}{\rho_l} + \frac{u}{\rho_v}, \tag{7}$$

where $\rho_l$ and $\rho_v$ are the density of the liquid and vapor, and $u$ is the vapor quality which is the percentage of mass that is vapor [4].

## 2.5 Single and Two Phase Flow

The working fluid in the micro-channel can be either single phase (which consists of exclusively liquid coolant as the working fluid) or two phase (which consists of both liquid and vapor).

When the power density is too high so that the liquid absorbs too much heat and its temperature increases dramatically, part of the liquid will become vapor and form two phase flow. The two phase flow exhibits different patterns. Figure 3a–f shows the two phase flow patterns in horizontal channels. When the flow rate is low, the flow usually exhibits bubbly (Fig. 3a) or plug pattern (Fig. 3b), as the flow rate increases, the pattern becomes stratified (Fig. 3c) and wavy (Fig. 3d), and finally slug (Fig. 3e) and annular (Fig. 3f) [16, 17].

The evaporation process in a channel is as Fig. 3g shows. As the single phase liquid absorbs heat so that the temperature increases to the evaporation point, small bubbles appear. When the fluid continues to absorb heat along the channel, plug and slug flows appear. The flow becomes waved and annular in the end.

Figure 4 compares the cooling effectiveness of single and two phase flows. It plots the solid temperature at the micro-channel outlet location $T_w$ versus the footprint power density $q_a$ for both single and two phase flows at same pumping power [18]. It shows that two phase flow achieves lower solid temperature than single phase flow, which indicates that two phase flow has higher cooling effectiveness than single phase flow.

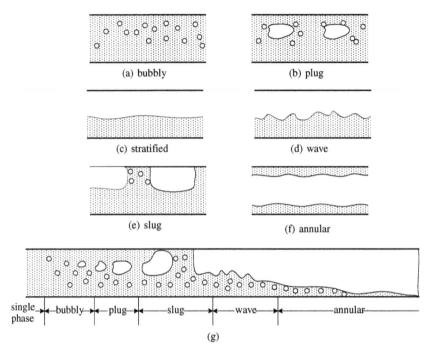

**Fig. 3.** (a)–(f) Two phase flow patterns, (g) evaporation process in a channel.

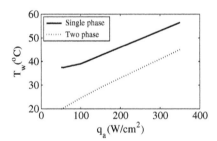

**Fig. 4.** Comparison of single and two phase flow.

## 2.6 Laminar and Turbulent Flow

The flow inside micro-channels can be laminar, turbulent, or transitional. Figures 5–7 show these three types of patterns. Laminar flow (Fig. 5) occurs when fluid flows in parallel layers, with no disruption between the layers. That is, the pathlines of different particles are parallel. It generally happens in small channels and low flow velocities. In turbulent flow (as shown in Fig. 6),

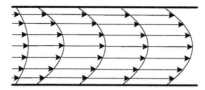

**Fig. 5.** Laminar flow pattern.

**Fig. 6.** Turbulent flow pattern.

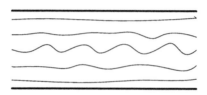

**Fig. 7.** Transitional flow pattern.

vortices and eddies appear, and make the flow unpredictable. Turbulent flow generally happens at high flow rates and larger channels. Transitional flow (Fig. 7) is a mixture of laminar and turbulent flow, with turbulence in the center of the channel, and laminar flow near the edges.

Usually, Reynolds number is used to predict the type of flow (whether laminar, turbulent, or transitional) in straight channels. For example, as [11] shows, usually:

When $Re < 2100$, it is laminar flow; when $2100 < Re < 4000$, it is transitional flow; when $Re > 4000$, it is turbulent flow.

When the channel involves more complex structure, the fluid exhibits more complicated behavior. Figure 8 shows an example of otherwise laminar flow in straight channels in a micro-channel with bends. When fluid enters a channel, it firstly subjects to a flow development process and after traveling some distance downstream, it becomes fully developed laminar flow. Then, when the flow comes across a bend, it becomes turbulent/developing around

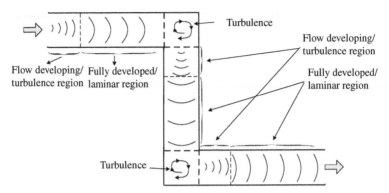

**Fig. 8.** Fluid in micro-channel with bends.

the corner and settles down after traveling some distance downstream into laminar fully developed flow again [19].

## 2.7 Fanning Friction Factor

Different types of flow (laminar/turbulent flow, fully developed/developing flow, etc.) behave in different manners in terms of their fanning frictional energy loss while flowing. Many works have been done to characterize the friction factor $f_r$ of different flow conditions and micro-channel geometries [20–22]. Table 1 summarizes some equations for calculating the friction factors for single phase laminar and turbulent flow. In the table, $\beta = b/d$ is the aspect ratio, where $b$ and $d$ are the width and height of micro-channel.

**Table 1** Summary of friction factors for single phase laminar and turbulent flow in rectangular micro-channels.

| Flow Type | Equation for Computing Friction Factor |
|-----------|----------------------------------------|
| Laminar | $f_r Re = 24(1 - 1.3553\beta + 1.9467\beta^2 - 1.7012\beta^3 \\ \quad\quad + 0.9564\beta^4 - 0.2537\beta^5)$ [3] |
|  | $f_r Re = 4.70 + 19.64\left(\frac{\beta^2+1}{(\beta+1)^2}\right)$ [21] |
| Turbulent | $f_r = \lambda Re^\sigma$ |
|  | $\lambda = 0.079, \sigma = -0.25$ [3] |
|  | $\lambda = 0.093 + \frac{1.016}{\beta}, \sigma = -0.268 - \frac{0.319}{\beta}$ [22] |

## 3. DESIGN CONSIDERATIONS OF MICRO-CHANNELS IN 3D-ICS

As shown in Fig. 1, each tier of 3D-IC contains an active silicon layer and silicon substrate. The micro-channels are placed horizontally in the silicon substrate. Through-silicon-vias (TSVs) such as power/ground TSV, signal TSV, etc., are used to enable communication between layers and delivery of power and ground. Figure 9 shows a possible configuration of micro-channels and TSVs in the silicon substrate of 3D-IC [23, 24]. In each 3D-IC tier, micro-channels are etched in the inter-layer region (silicon substrate). Fluidic channels (fluidic TSVs) go through all the tiers and delivers coolant to micro-channels. TSVs also go through the silicon substrate vertically to deliver signal, power, and ground.

Though the micro-channel heat sink is capable of achieving good cooling performance, many problems need to be addressed when designing the micro-channel infrastructure for cooling 3D-IC so as to ensure the reliability of the chip and improve the effectiveness of the micro-channel.

### 3.1 Non-Uniform Power Profile

The underlying heat dissipated in each active silicon layer exhibits great non-uniformity [3, 25]. For instance, Fig. 10a shows a typical power profile of a processor [26]. This power profile is generated by simulating a high performance out-of-order processor. The detailed configuration of the processor is described in [26]. To obtain the power profiles for the processor, Zhang and Srivastava [26] simulated SPEC 2000 CPU benchmarks [27],

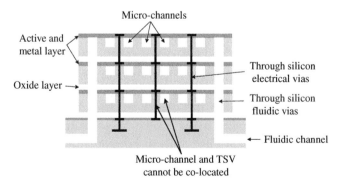

**Fig. 9.** Cross section of 3D-IC with micro-channels and TSVs.

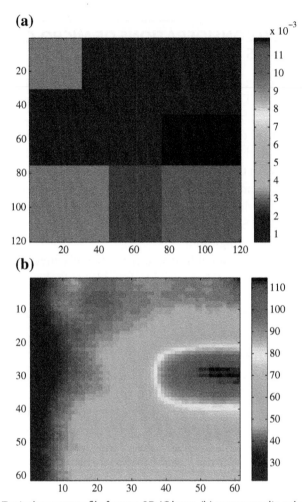

**Fig. 10.** (a) Typical power profile for one 3D-IC layer, (b) corresponding thermal profile.

and for each benchmark, they simulated a representative $250M$ instructions and sampled the chip power dissipation values using uniform time intervals. Figure 10a shows a sample power profile achieved in the simulation. The power profile in Fig. 10a shows that the power density is generally high in instruction window, and much lower in other regions such as caches. Such non–uniformity in power profile results in hotspots in thermal profiles as Fig. 10b shows. Therefore, when designing micro–channel heat sink infrastructure, one should account for this non–uniformity in thermal and power profiles. Simply minimizing the total resistance of the chip while failing to consider the non-uniformity of the power profile will lead to sub-optimal design. For example, conventional approaches for micro-channel designs spread the entire surface to be cooled with channels [2, 5]. This approach, though helps reducing the peak temperature around the hotspot region, overcools areas that are already sufficiently cool. This is wasteful from the point of view of pumping power.

## 3.2 TSV Constraint

3D-ICs impose significant constraints on how and where the micro-channels could be located due to the presence of TSVs, which allow different layers to communicate. Micro-channels are allocated in the interlayer bulk silicon regions. TSVs also exist in this region, causing a resource conflict. A 3D-IC usually contains thousands of TSVs which are incorporated with clustered or distributed topologies [28, 29]. These TSVs form obstacles to the micro-channels since the micro-channels cannot be placed at the locations of TSVs. Therefore the presence of TSVs limits the available spaces for micro-channels, and designing the micro-channel infrastructure should take this fact into consideration.

## 3.3 Thermal Stress

The TSV fill materials are usually different from silicon. For example, copper has low resistivity and is therefore widely used as the material for TSV fill. Because the annealing temperature is usually much higher than the operating temperature, thermal stress will appear in silicon substrate and TSV after cooling down to room temperature due to the thermal expansion mismatch between copper and silicon [30, 31]. This thermal stress might result in structure deformation, thereby causing reliability problems such as cracking.

Moreover, as shown in [30, 32], thermal stress also influences carrier mobilities significantly, hence changing the gate delay. As Eqn (8) shows, the mobility change ($\Delta\alpha$) due to thermal stress depends on the intensity and orientation of applied stress, and the type of transistor [30, 32, 33]:

$$\frac{\Delta\alpha_j}{\alpha_j} = \Pi \sum_{i \in \mathrm{TSV}(j)} s_{rr,ij} \times \psi(\omega_{ij}). \qquad (8)$$

Here $\Pi$ is a coefficient indicating the sensitivity of mobility change to applied stress, TSV($j$) represents the set of TSVs that influence gate $j$, $s_{rr,ij}$ is the radial stress applied on gate $j$ caused by $i$th TSV, $\omega_{ij}$ is defined as the angle between the line connecting the center of TSV $i$ and transistor $j$, and the channel direction of transistor $j$ (as Eqn (9) and Fig. 11 show). $\psi$ is an orientation

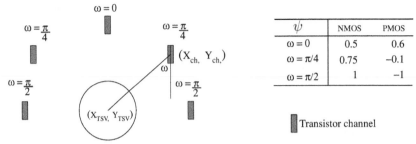

**Fig. 11.** Channel position with respect to TSV ($\omega$), and values of orientation coefficient $\psi(\omega)$.

coefficient whose value depends on $\omega$ and the type of transistor. Figure 11 gives the value of $\psi$ for different types of transistors and different values of $\omega$:

$$\omega_{ij} = \arctan \frac{|X_{TSV,i} - X_{ch,j}|}{|Y_{TSV,i} - Y_{ch,j}|}, \qquad (9)$$

where $(X_{TSV,i}, Y_{TSV,i})$ and $(X_{ch,j}, Y_{ch,j})$ indicate the coordinates of TSV center and transistor channel.

As shown in Eqn (8), thermal stress influences carrier mobilities, hence changing the gate delay. Therefore, if the gates on critical paths are allocated near TSVs (basically regions with high thermal stress), timing violation might occur.

The existence of micro-channels which influences the temperature around TSVs will influence the thermal stress (either increase or decrease thermal stress), thereby changing the mechanical reliability analysis and timing analysis in the 3D-IC with TSVs. For example, Fig. 12 shows the thermal stress inside and surrounding a TSV at different thermal conditions. Figure 12a depicts the thermal stress when chip temperature is 100 °C and annealing temperature (which is basically the stress free reference temperature) is 250 °C. The figure shows that large thermal stress (up to 490 MPa) appears surrounding the TSV. Figure 12b depicts the thermal stress when the chip temperature is 50 °C. In this case (where chip temperature is 50 °C), the overall thermal stress is increased (compared with the previous case where chip temperature is 100 °C), and the maximum thermal stress reaches up to 670 MPa. Such phenomenon indicates that reduction in chip temperature results in an increase in thermal stress. Hence the existence of micro-channels, which generally reduces chip temperature, may increase the thermal stress caused by TSVs. Such phenomenon should be considered when designing the micro-channel infrastructure.

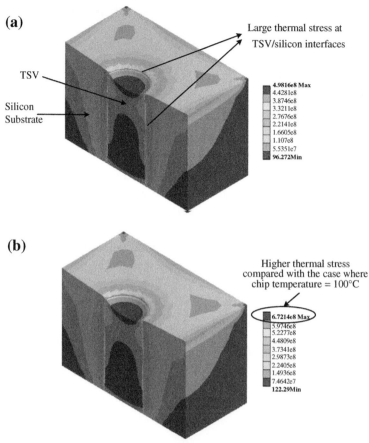

**Fig. 12.** Thermal stress inside and surrounding TSV (a) when chip temperature is 100 °C, (b) when chip temperature is 50 °C (assuming stress free temperature is 250 °C).

Moreover, if micro–channels are placed too close to the TSVs, the silicon walls between the TSV and micro–channel will be more likely to crack because this wall is thin. These facts further limit the locations of micro–channels.

## 4. 3D-IC WITH MICRO-CHANNELS: SIMPLIFIED MODELING

The thermal and hydrodynamic characteristic of fluids in micro–channels are basically governed by the conservation equations in Eqn (1). Computational fluid dynamic (CFD) tools are developed to simulate the thermal and hydraulic behavior of fluids. For example, the ANSYS CFX [34] can be used to simulate 3D–IC integrated with interlayer micro–channel heat sinks. These tools usually use finite volume approach based on the conservation equations, which are very slow. Therefore, researchers have developed approaches for modeling the 3D–IC liquid cooling with less computational complexity while still offering reasonably good accuracy.

## 4.1 Thermal Modeling

### 4.1.1 Bulk Thermal Resistance

Based on the heat transfer phenomenon described in Section 2.1, two types of thermal resistances are used to model the heat removal through micro–channels: convective resistance $R_{conv}$ and heat resistance $R_{heat}$ [2].

Tuckerman and Pease [2] and Knight et al. [5] assume the heat flux generated in the solid layer is constant and therefore the temperature difference between the solid heat source and the fluid is the same along any plane in the fluid direction (as shown in Fig. 13). Therefore, the heat transfer rate from the solid channel sidewalls to the fluid equals the power generated in the solid active layer:

$$q = (T_{w,i} - T_{f,i})/R_{conv} = (T_{w,o} - T_{f,o})/R_{conv}, \qquad (10)$$

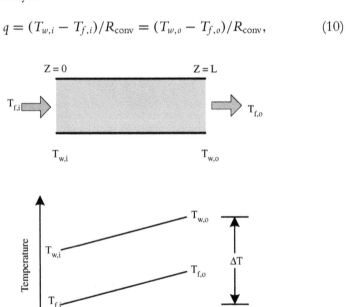

**Fig. 13.** Temperature variation in micro-channel fluid and solid under constant heat flux.

where $q$ is the power dissipation in the solid, $T_{f,i}$ is the average fluid temperature at the micro-channel inlets (that is, at the location $z = 0$), and $T_{w,i}$ is the average solid temperature at plane $z = 0$. Similarly, $T_{w,o}$ and $T_{f,o}$ are the average solid and fluid temperature at the micro-channel outlet plane $z = L$ (as shown in Fig. 13).

The convective resistance $R_{\mathrm{conv}}$ represents the heat convection from micro-channel sidewalls to the fluid. It can be estimated by:

$$R_{\mathrm{conv}} = 1/hA, \tag{11}$$

where $h$ is the heat transfer coefficient which could be estimated by Eqn (4), and $A$ is the total surface area available for heat transfer in micro-channels, whose value depends on the structure of micro-channels [2, 5].

Then this same amount of power $q$ (in Eqn (10)) is carried away through the moving fluid which results in temperature increase downstream:

$$q = (T_{f,o} - T_{f,i})/R_{\mathrm{heat}}. \tag{12}$$

Here $T_{f,o}$ and $T_{f,i}$ are the average outlet and inlet temperature of the fluid, and $R_{\mathrm{heat}}$ is the heat resistance calculated by:

$$R_{\mathrm{heat}} = 1/\rho F C_v. \tag{13}$$

So the overall thermal resistance $R_{\mathrm{total}}$ becomes:

$$R_{\mathrm{total}} = \frac{\Delta T}{q} = \frac{T_{w,o} - T_{f,i}}{q} = \frac{1}{hA} + \frac{1}{\rho F C_v}. \tag{14}$$

Generally, smaller thermal resistance indicates better micro-channel cooling performance.

### 4.1.2 RC Network

The previous model uses a bulk thermal resistance to characterize the heat transfer associated with micro-channels. This bulk thermal resistance model, though gives an overall idea of the heat removal capability of micro-channels, fails to consider the non-uniform nature of power dissipation in 3D-ICs. In addition, it does not allow one to extract the chip thermal profile from the power profile. The work in [10, 35] tried to address this problem, and uses a HOTSPOT-like [36] distributed resistance-capacitance RC network for thermal modeling of 3D-IC with micro-channels.

Basically their approach is first dividing the 3D-IC into small grids, and then use numerical approach to approximate the PDE-based energy conservation equation in Eqn (1) with ordinary differential equations (ODE). Assuming the size of each grid is $\Delta x \times \Delta y \times \Delta z$, let $T_{i,j,k}$ be the temperature at the $i$th/$j$th/$k$th grid in $x$/$y$/$z$ direction, for a grid in the fluid domain in micro-channels, the numerical approximation for this grid works as follows:

$$C_v \frac{dT}{dt} + \nabla \cdot (-k_f \nabla T) + C_v \vec{V} \cdot \nabla T = \dot{q}.$$

$$
\begin{aligned}
\Rightarrow C_v \Delta x \Delta y \Delta z \frac{dT}{dt} \;-\; & k_{f,x} \frac{T_{i+1,j,k} - 2T_{i,j,k} + T_{i-1,j,k}}{\Delta x^2} \\
-\; & k_{f,y} \frac{T_{i,j+1,k} - 2T_{i,j,k} + T_{i,j-1,k}}{\Delta y^2} \\
-\; & k_{f,z} \frac{T_{i,j,k+1} - 2T_{i,j,k} + T_{i,j,k-1}}{\Delta z^2} \\
+\; & C_v V_{\text{avg},z} \Delta x \Delta y (T_{s2} - T_{s1}) = \dot{q}\Delta x \Delta y \Delta z,
\end{aligned}
\tag{15}
$$

where $k_{f,x}, k_{f,y}, k_{f,z}$ are the fluid thermal conductivity in $x, y, z$ direction, $V_{avg,z}$ is the average fluid velocity in the fluid direction $z$. $T_{s1}$ and $T_{s2}$ are the surface temperature at the front and back of this grid (as shown in Fig. 14a), and can be approximated as $T_{s1} = (T_{i,j,k-1} + T_{i,j,k})/2$ and $T_{s2} = (T_{i,j,k} + T_{i,j,k+1})/2$.

Based on the numerical approximation, each grid in the micro-channel can be modeled as an RC circuit as Fig. 14b shows. In the RC circuit, the heat conduction in the fluid and convection between the fluid and sidewall is represented by resistance, and the heat transfer carried by the moving flow is represented by a voltage controlled current source. For grids in solid domain, similar model can be derived as [10] illustrates. By combining the RC circuit of each grid, the 3D-IC with integrated micro-channel heat sink can be represented by a distributed RC network (which is equivalent to a system of ODEs). Given the power profile, the temperature profile at any given time could be estimated by numerically solving the system of ODEs using Euler's method, etc.

### 4.1.3 Resistive Network

The work in [37] uses a resistive network to model the steady state thermal behavior of 3D-IC with micro-channels. Similarly as [10], they discretize the micro-channel into sections along the fluid direction $z$. Let $R_{\text{conv}}$ be the convective resistance and $R_{\text{heat}}$ be the heat resistance of one of these sections. The formulations estimating $R_{\text{conv}}$ and $R_{\text{heat}}$ for each section are

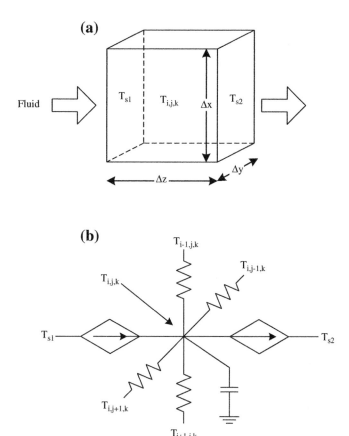

**Fig. 14.** (a) Grid $(i, j, k)$ and (b) its equivalent RC circuit.

similar as Section 4.1.1. Therefore, each micro-channel section is represented by a thermal resistive network composed of convective resistance $R_{conv}$ and heat resistance $R_{heat}$ as Fig. 15 shows.

Similar as the HOTSPOT approach [36], the 3D–IC is partitioned into fine grids and represented as resistive network. The micro-channel resistance network described above and the 3D–IC resistive network are combined to generate a unified model that captures the steady state thermal behavior of 3D–ICs with liquid cooling. For a given resistive network and the power profile, the temperature at each grid can be estimated using Eqn (16), where $H$ is a matrix decided by the thermal resistance network, $Q$ is a matrix representing the chip power profile, and $T$ represents the temperature profile:

$$H \cdot T = Q. \tag{16}$$

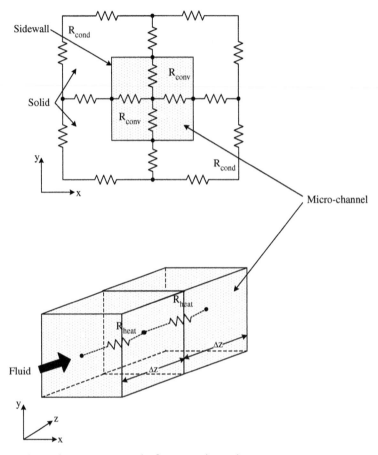

**Fig. 15.** Thermal resistive network of a micro-channel section.

Table 2 summarizes different thermal modeling approaches for 3D-IC with micro-channels.

## 4.2 Hydrodynamic Modeling

The power used by micro-channels for performing chip cooling comes from the work done by the fluid pump to push the coolant fluid through micro-channels. The pumping power $Q_{pump}$ is decided by the pressure drop and volumetric flow rate of micro-channels:

$$Q_{pump} = \sum_{ch=1}^{N} F_{ch} \Delta P_{ch}. \tag{17}$$

**Table 2** Summary of thermal modeling approaches for 3D-IC with micro-channels.

| Approach | Description | Governing Equations | Advantages | Disadvantages |
|---|---|---|---|---|
| Bulk resistance | Uses two bulk thermal resistances $R_{conv}$ and $R_{heat}$ to model heat transfer through one micro-channel | Equations (11), (13), and (14) | 1. Gives overall idea of micro-channel heat removal capability;<br>2. Easy to calculate | 1. Fails to consider non-uniformity in power profile;<br>2. Unable to generate fine-grained thermal profile |
| RC network | Uses distributed RC network; Resistance: heat transfer, capacitance: heat storage | Equation (16) | Generates transient 3D-IC thermal profile | Computationally complex |
| Resistive network | Uses distributed resistance network to model steady state thermal behavior | Equation (16) | 1. Generates steady state thermal profile;<br>2. Easier to compute than RC network | Unable to model transient thermal behavior |

Assuming there are a total of $N$ micro-channels, here $\Delta P_{ch}$ and $F_{ch}$ are the pressure drop and volumetric flow rate of the $ch$th micro-channel. Pressure drop and fluid flow rate are interdependent and related to other micro-channel parameters such as length and width. It also depends on the shape of the channel and the property of fluid inside the channel. Many papers investigated the correlation between pressure drop and flow rate [5, 19, 20, 38]. In the following subsections, we will introduce pressure drop modeling for different micro-channel structures.

### 4.2.1 Pressure Drop in Straight Micro-Channels

Kandlikar et al. [38] models the pressure drop in single phase flows in straight micro-/mini-channels. The pressure drop in a straight channel is usually modeled as:

$$\Delta P = \frac{2\gamma\mu VL}{D_h^2} = \frac{2f_r\rho V^2 L}{D_h},\tag{18}$$

where $V$ is the fluid velocity, $L$ is the length of the micro-channel, $D_h$ is the hydraulic diameter. Here $\mu$ is fluid viscosity and $\gamma$ is determined by the micro-channel dimension (given in [5]), $f_r$ is the friction factor whose expression depends on the fluid condition (single/two phase and laminar/turbulent) and the shape of the cross section of the micro-channel [20–22].

### 4.2.2 Pressure Drop in Bended Micro-Channels

Consider the channel structure shown in Fig. 8. The existence of a bend causes a change in the flow properties which impact the cooling effectiveness and pressure drop. An otherwise fully developed laminar flow in the straight part of the channel, when comes across a 90° bend becomes turbulent/developing around the corner and settles down after traveling some distance downstream into laminar fully developed again (see Fig. 8). So a channel with bends has three distinct regions, (1) fully developed laminar flow region, (2) the bend corner, and (3) the developing/turbulent region after the bend [19, 38]. The length of flow developing region is [39]:

$$L_d = (0.06 + 0.07\beta - 0.04\beta^2)Re \cdot D_h,\tag{19}$$

where $\beta = b/d$ is the channel aspect ratio, and $Re$ is the Reynolds number.

The rectangular bend impacts the pressure drop. Due to the presence of bends, the pressure drop in the channel is greater than an equivalent straight channel with the same length and cross section. The total pressure drop in a channel with bends is the sum of the pressure drop in the three regions described above (which finally depends on how many bends the channel has). Assuming $L$ is the total channel length, and $m$ is the bend count. Hence $m \cdot L_d$ is the total length that has developing/turbulent flow and $m \cdot b$ is the total length attributed to corners (see Fig. 8). Hence the effective channel length attributed to fully developed laminar flow is $L - m \cdot L_d - m \cdot b$. The pressure drop in the channel is the sum of the pressure drop in each of these regions.

### Pressure Drop in Fully Developed Laminar Region

The total pressure drop in fully developed laminar region is [5]:

$$\Delta P_f = \frac{2\gamma\mu(L - m \cdot L_d - m \cdot b)V}{D_h^2} = \frac{2\gamma\mu L_f V}{D_h^2}. \tag{20}$$

Here $L_f = L - m \cdot L_d - m \cdot b$ is the total length of the fully developed laminar region which is explained above, the other parameters are the same as in Eqn (18).

### Pressure Drop in Flow Developing Region

The pressure drop in each flow developing region is [20]:

$$\Delta p_d = \frac{2\mu V}{D_h^2} \int_0^{L_d} \theta(z)dz. \tag{21}$$

Here $\theta(z)$ is given by $\theta(z) = 3.44\sqrt{(Re \cdot D_h)/z}$, where $z$ is the distance from the entrance of developing region in the flow direction. Assuming there are a total of $m$ corners in a given micro-channel, so there are $m$ developing regions with the same length $L_d$ in this channel. By putting the expression of $\theta(z)$ and $L_d$ into Eqn (21) and solving the integration, we can get the total pressure drop of the developing region in this micro-channel:

$$\Delta P_d = m\Delta p_d = m\rho K_d V^2, \tag{22}$$

where $K_d = 13.76(0.06 + 0.07\beta - 0.04\beta^2)^{\frac{1}{2}}$ is a constant associated with the aspect ratio $\beta$. Please refer to [20, 38] for details.

### Pressure Drop in Corner Region

The total pressure drop at all the 90° bend in a micro-channel is decided by:

$$\Delta P_{90°} = m\Delta p_{90°} = m\frac{\rho}{2}K_{90°}V^2, \tag{23}$$

where $m$ is the number of corners in the channel, $\Delta p_{90°}$ is the pressure drop at each bend corner, and $K_{90°}$ is the pressure loss coefficient for 90° bend whose value can be found in [38].

## Total Pumping Power

The total pressure drop in a micro-channel with bends is the sum of the pressure drop in the three types of regions discussed above:

$$\Delta P = \Delta P_d + \Delta P_f + \Delta P_{90°}$$
$$= \frac{2\gamma \mu L_f}{D_h^2} V + m(K_d + \frac{K_{90°}}{2})\rho V^2. \tag{24}$$

From Eqn (24), the total pressure drop of a micro-channel is a quadratic function of the fluid velocity $V$. For a given pressure difference applied on a micro-channel, we can calculate the associated fluid velocity by solving Eqn (24). With the fluid velocity, we can then estimate the fluid flow rate $F$, and thus estimate the thermal resistance and pumping power for this channel. Hence the pumping power as well as cooling effectiveness of micro-channels with bends is a function of (1) number of bends, (2) location of bends, and (3) pressure drop across the channel.

Comparing Eqns (18) and (24), due to the presence of bends, if the same pressure drop is applied on a straight and a bended micro-channel of the same length and cross section, the bended channel will have lower fluid velocity, which leads to a lower cooling capability. Therefore, to provide sufficient cooling, we will need to increase the overall pressure drop that the pump delivers, which results in increase of pumping power. But bends allow for better coverage of hotspots in the presence of TSVs.

Table 3 gives a summary of hydrodynamic modeling of fluids in different micro-channel structures.

**Table 3** Summary of hydrodynamic modeling for fluids in different micro-channel structures.

| Structure | Regions | Pressure Drop Model |
|---|---|---|
| Straight | Developing | $\Delta p_d = \rho K_d V^2$ |
| | Fully developed laminar | $\Delta p_f = \frac{2f_r \rho V^2 L_f}{D_h} = \frac{2\gamma \mu L_f V}{D_h^2}$ |
| Bended | Developing | $\Delta p_d = \rho K_d V^2$ |
| | Fully developed laminar | $\Delta p_f = \frac{2f_r \rho V^2 L_f}{D_h} = \frac{2\gamma \mu L_f V}{D_h^2}$ |
| | Corner | $\Delta p_{90°} = \frac{\rho}{2} K_{90°} V^2$ |

$K_d, K_{90°}$: pressure loss coefficient; $V$: fluid velocity; $L_f$: length of fully developed region; $f_r$: fraction factor; $D_h$: hydraulic diameter; $\rho$, $\mu$: fluid density, viscosity.

## 4.3 Cooling Effectiveness with Respect to Micro-Channel Parameters

From the models described above, there are several parameters that could be used to control the cooling performance of micro-channel heat sinks.

### 4.3.1 Flow Rate

The fluid flow rate $F$ (explained in Section 2.4) influences the micro-channel cooling performance by changing the heat resistance $R_{heat}$ (given in Eqn (13)). From Section 4.1.1 we can see, when the fluid flow rate increases, $R_{heat}$ reduces, which results in better cooling performance. Figure 16a plots the maximum solid temperature at micro-channel output plane versus the 3D chip power density for different flow rates. As the flow rate increases, the maximum solid temperature decreases which means better micro-channel cooling performance. However, increase in fluid flow rate also leads to increase in pumping power (see Eqns (17), (18), and (24)). Therefore, increase in flow rate results in improved cooling performance at a cost of increased pumping power consumption.

### 4.3.2 Number of Micro-Channels

The distribution of micro-channels will also influence the cooling performance significantly. For example, for straight and uniformly distributed micro-channels, increase in the number of channels $N$ helps improving the coverage of the cooling system. Therefore, although more micro-channels are used, a lower flow rate per channel could be used to provide enough cooling and thereby reduce the pumping power consumption. Figure 16b plots the pumping power required to cool the 3D-IC below 80 °C versus the power density for different number of uniformly distributed micro-channels. For each micro-channel design (with different number of channels), we use the minimum pressure drop $\Delta P$ that can provide enough cooling for the 3D-IC. Generally, increase in number of channels could reduce the required flow rate and thereby reduce the pumping power. However, when the power density is low, increase in the number of channels does not help.

### 4.3.3 Structure of Micro-Channels

The structure of micro-channels also influences their cooling effectiveness. Comparing Eqns (18) and (24), due to the presence of bends, if the same pressure drop is applied on a straight and a bended micro-channel of the same length, the bended channel will have lower fluid velocity, which leads

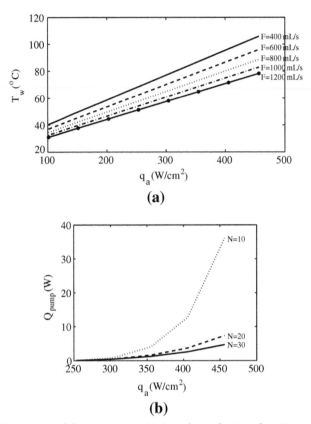

**Fig. 16.** (a) Maximum solid temperature at micro-channel output location versus power density for different flow rates, (b) pumping power versus power density for different number of micro-channels.

to a lower average cooling capability. But bends allow for better coverage of hotspot in the presence of TSVs. Therefore the number and location of bends should be carefully selected.

## 5. MICRO-CHANNEL DESIGN CHALLENGES

### 5.1 Micro-Channel Modeling

While some works have investigated and proposed thermal and hydrody-namic modeling approaches for 3D–IC with micro–channels, basically, these approaches target at some specific structures such as straight rectangular or micro-pin-fin structures [10, 35]. Better modeling approaches should be investigated. The models should: (a) capture both the transit and steady state thermal behaviors and (b) handle any complex micro-channel infrastructures (which will be discussed below).

## 5.2 Micro-Channel Infrastructure Design

Another challenge is micro-channel infrastructure design. Many works have been done to optimize the cooling effectiveness of the micro-channels. These works can be classified into three directions: shape optimization of straight micro-channels, complex micro-channel infrastructure, and hotspot-optimized micro-channel structure.

### 5.2.1 Straight Micro-Channel Shape Optimization

Some works try to improve the cooling effectiveness of straight micro-channels by controlling their dimensional parameters such as channel width, height, fin thickness, etc. [2, 5, 40, 41]. Most of these approaches try to optimize the micro-channel cooling effectiveness by minimizing the overall thermal resistance. However, such approaches fail to consider the non-uniform nature of the underlying power profile of 3D-IC, and therefore might result in sub-optimal design.

### 5.2.2 Complex Micro-Channel Infrastructures

Some other works try to investigate more complex micro-channel infrastructures which could achieve better cooling effectiveness than straight rectangular micro-channels. These structures include micro-channels with trapezoidal cross sections [42, 43], cross-linked micro-channels [44], micro-pin-fins [45–47], tree-shaped and serpentine structures [19, 48–50], and four-port micro-channel structure [51, 52].

### Cross-Linked Micro-Channels

Due to the non-uniformity of thermal profile on chip, the fluid temperature inside different micro-channels is also non-uniform. Such non-uniformity causes inefficiency. In cross-linked micro-channel structure, micro-channels are connected as Fig. 17 shows [44]. The cross link between micro-channels allows fluid exchange between different channels. The fluid exchange results in more uniform fluid temperature distribution among micro-channels, thereby improving micro-channel cooling effectiveness.

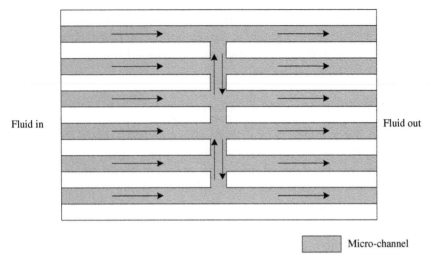

Fluid in                                                                        Fluid out

Micro-channel

**Fig. 17.** Top view of cross-linked micro-channels.

## Micro-Pin-Fins

Some works investigate micro-pin-fins instead of micro-channels [45–47]. As Fig. 18 shows, micro-pin-fins are used for connection between different 3D-IC layers. Such structure might result in larger area of heat transfer, which helps improving the cooling effectiveness. The pin-fin in-line structure (as Fig. 18a depicts) and pin-fin staggered structure (as Fig. 18b depicts) are two common micro-pin-fin structures. Brunschwiler et al.[53] compares the cooling effectiveness of pin-fin in-line, pin-fin staggered, and straight micro-channel structures, and found that the pin-fin in-line structure outperforms the other two structures.

When applying micro-pin-fin structure to 3D-IC cooling, a few considerations should be accounted for. First, since the 3D-IC stacks several layers vertically, the diameter and pitch of micro-pin-fins should be carefully decided to avoid mechanical reliability problems. Second, since the micro-pin-fins are used to enable connection and communications between different 3D-IC layers, TSVs are basically contained inside these pin-fins. Hence when allocating or designing micro-pin-fins, one should consider their impact on TSVs or the constraints imposed by TSVs.

## Tree-Shaped and Serpentine Structure

Tree and serpentine structured micro-channels are also explored [19, 48–50]. Figure 19 shows two types of tree-shaped micro-channel structures, while

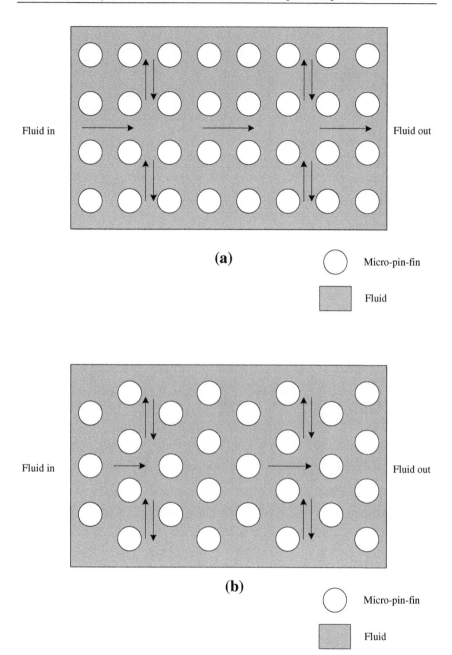

**Fig. 18.** Top view of micro-pin-fin structure: (a) pin-fin in-line, (b) pin-fin staggered.

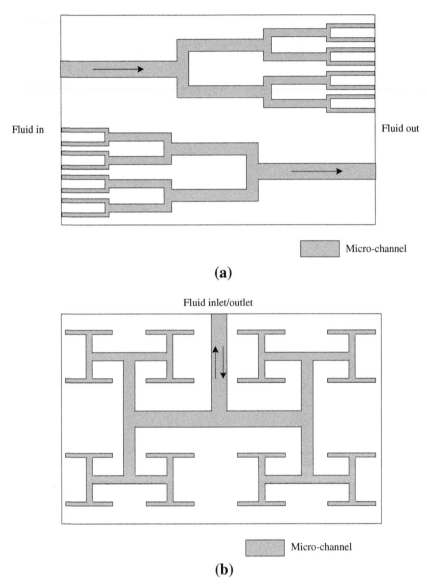

**Fig. 19.** Two tree-shaped micro-channel structures.

Fig. 20 gives a serpentine structured micro-channel. As Fig. 19a shows, the first tree-shaped structure can provide different cooling capability to different on-chip locations flexibly. On the other hand, the second tree-shaped micro-channel structure (Fig. 19b) and the serpentine structure (Fig. 20) distribute coolant more uniformly, but their advantage is that only one or

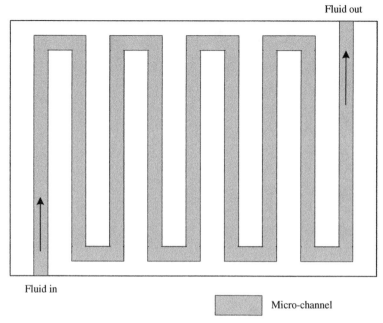

**Fig. 20.** Serpentine structured micro-channel.

two inlet/outlet ports are required in each micro-channel layer, which is easier to manufacture and manage.

## Four-Port Micro-Channel/Micro-Pin-Fin Structure

Brunschwiler et al. [51] proposed four-port micro-channel/micro-pin-fin structure (as Fig. 21 depicts) which can enhance the total volumetric fluid flow rate. Compared with the two-port structure, the fluid velocity at the four corners is drastically enhanced due to the short fluid path from inlets to outlets and therefore improve the cooling effectiveness at the corner regions. However, the drawback of this structure is that the fluid velocity in the center is low. As a result, hotspots may appear in the center region.

### 5.2.3 Hotspot-Optimized Micro-Channel Infrastructure

To account for the non-uniform power profile of 3D-IC, Shi et al. [37] Shi and Srivastava [54] Brunschwiler et al. [51] developed hotspot-optimized micro-channel infrastructure which allocates or designs the micro-channel according to the variations in power and thermal profiles.

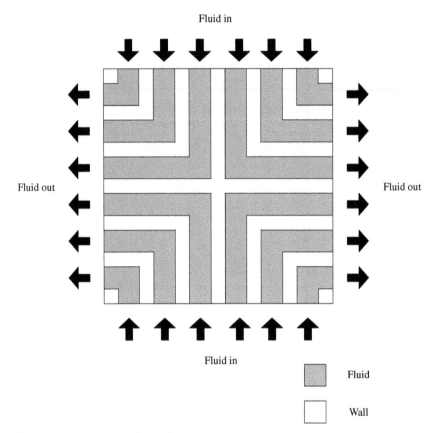

**Fig. 21.** Four-port micro-channel structure.

## Non-Uniformly Distributed Straight Micro-Channels

Shi et al. [37] investigate *non-uniform* allocation of straight micro-channels. It is well known that CPU exhibits significant variation in power and thermal profiles. Spreading channels uniformly across the whole chip, though helps in reducing the peak temperature at hotspot regions, overcools areas that are already sufficiently cool. This is wasteful from the point of view of pumping power. Moreover, as explained in Section 3.2, the presence of TSVs that connect signals, power between layers constraints the uniform spreading of micro-channels by limiting the areas where channels could be placed.

The work in [37] places micro-channels in high heat density areas, and saves cooling power compared with *uniform* micro-channel structure. As Fig. 22a and b shows, the conventional straight micro-channel design spreads micro-channels all over the chip (except in those regions that contain TSVs

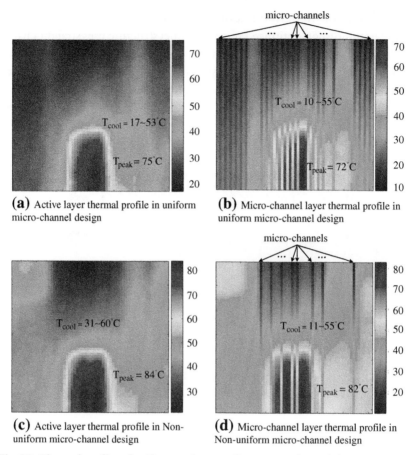

**Fig. 22.** Thermal profiles of uniform and non-uniform micro-channel designs.

or other structures). In this conventional *uniform* micro-channel structure, in most on-chip locations (other than the hotspots), the temperature is low (from 10 °C to 50 °C). So placing micro-channels in these locations is not necessary. On the contrary, the *non-uniform* micro-channel design proposed in [37] (shown in Fig. 22c and d) only places micro-channels in hotspot regions. As we can see from Fig. 22d which is the temperature profile at micro-channel layer in *non-uniform* design, micro-channels on the left- and right-hand side of the chip (which are the cooler regions in *uniform* design) are very sparse. Although the temperature in this region is increased compared to the *uniform* design, it is still much lower than the temperature constraint. [37] reports that such *non-uniform* micro-channel distribution results in about 50% cooling power savings.

## Micro-Channel with Bended Configuration

The previous hotspot-optimized micro-channel structure uses straight channels that spread in areas that demand high cooling capacity. If the spatial distribution of micro-channels is unconstrained then such an approach results in the best cooling efficiency with the minimum cooling energy (power dissipated to pump the fluid). However 3D-ICs impose significant constraints

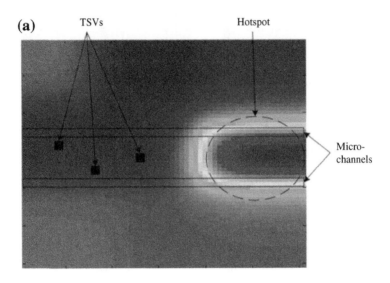

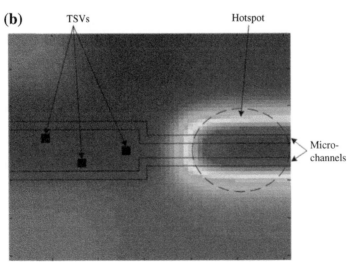

**Fig. 23.** Example of silicon layer thermal profile with TSV and (a) straight, (b) bended micro-channels.

on how and where the micro-channels could be located due to the presence of TSVs, which allow different layers to communicate. A 3D-IC usually contains thousands of TSVs which are incorporated with clustered or distributed topologies [29]. These TSVs form obstacles to the micro-channels since the channels cannot be placed at the locations of TSVs. Therefore the presence of TSVs prevents efficient distribution of straight micro-channels. This results in the following problems:

1. As illustrated in Fig. 23a, micro-channels would fail to reach thermally critical areas thereby resulting in thermal violations and hotspots.

2. To fix the thermal hotspots in areas where micro-channels cannot reach, we need to increase the fluid flow rate resulting in a significant increase in cooling energy.

To address this problem, Shi and Srivastava [54] investigated micro-channels with bended structure (as Fig. 23b depicts) to account for the distribution of hotspots and the constraints imposed by TSVs. As shown in Fig. 23b, with bended structure, the micro-channels can reach those TSV-blocked hotspot regions which straight micro-channels cannot reach. This results in better coverage of hotspots.

On the other hand, as Section 4.2 illustrates, due to the presence of bends, if the same pressure drop is applied on a straight and a bended micro-channel of the same length, the bended channel will have lower fluid velocity, which leads to a lower average cooling capability. Therefore, to provide sufficient

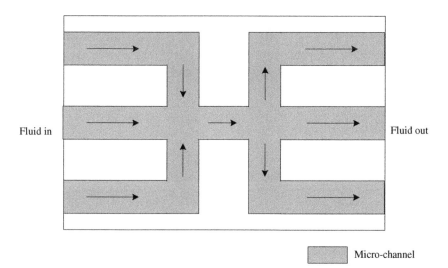

Micro-channel

**Fig. 24.** Fluid-focused micro-channel structure.

cooling, one might need to increase the overall pressure drop that the pump delivers, which results in increase of pumping power.

However bends allow for better coverage in the presence of TSVs. Therefore, the structure of bended micro–channels (especially the locations of bends) should be carefully selected so as to achieve maximum

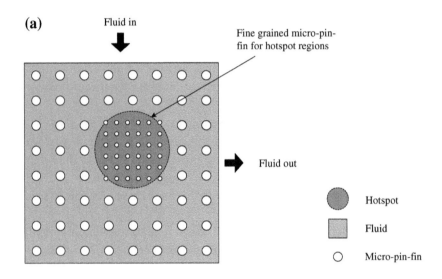

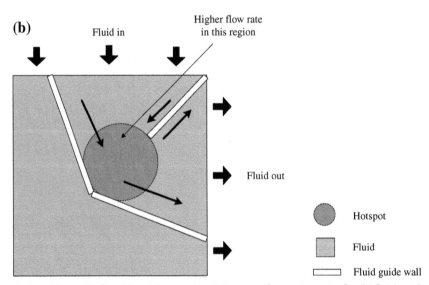

**Fig. 25.** Micro-pin-fin with guide structure (a) non-uniform micro-pin-fin, (b) fluid guide structure.

cooling efficiency. Shi and Srivastava [54] uses minimum cost flow-based approach to find the best bended micro-channel structure, and reports 13–87% cooling power savings compared with straight micro-channel infrastructure.

## Other Hotspot-Optimized Micro-Channel Configuration

Brunschwiler et al. [51] also investigated several hotspot-optimized approaches for micro-channel optimization including fluid-focused structure and micro-pin-fin with guide structure. In the fluid-focused structure (as shown in Fig. 24), the fluid is focused toward a hotspot location. Hence locally at the hot spot, the flow rate is increased, thereby improving the micro-channel cooling effectiveness at the hotspot region. We should also note that the structure needs to be balanced such that areas other than fluid-focused region (usually regions with lower heat density) still get sufficient cooling to stay below the maximum allowable temperature.

The micro-pin-fin with guide structure shown in Fig. 25 uses finer grained micro-pin-fins in hotspot regions to improve the cooling effectiveness of micro-pin-fins in that region (as Fig. 25a shows). Additionally, with guide structure (the guide walls as Fig. 25b) shows, fluid can be focused to the hotspot region so that the fluid at hotspot region has higher flow rate.

Table 4 summarizes different micro-channel infrastructures and their advantages.

**Table 4** Summary of different micro-channel infrastructure.

| Structure | Description | Advantages |
| --- | --- | --- |
| Straight | Figure 1 | Easy to manufacture and control |
| Cross-linked | Figure 17 | Enables more uniform fluid temperature distribution |
| Micro-pin-fin | Figure 18 | Larger area of heat transfer |
| Tree-shaped | Figure 19 | Provides different cooling capability to different on-chip locations |
| Serpentine | Figure 20 | Less fluid inlet/outlet ports, easier to manufacture and manage |
| Four-port | Figure 21 | Fluid velocity at corners is enhanced |
| Non-uniform channel | Figure 22 | Adapts to non-uniform power/thermal profiles with various hotspots |
| Bended-structure | Figure 23 | Overcomes constraints imposed by TSVs, better coverage of hotspots |
| Fluid-focused | Figure 24 | Provides better cooling to hotspots |
| Pin-fin with guide | Figure 25 | Provides better cooling to hotspots |

# 6. RUNTIME THERMAL MANAGEMENT USING MICRO-CHANNELS

Recently, the micro-channel heat sink is also adopted in dynamic thermal management (DTM) to control the runtime CPU performance and chip temperature by tuning the fluid flow rate through micro-channels [55–57].

The work in [55] proposed a DTM approach by controlling fluid flow rate through micro-channels. This enables us to change the configuration of cooling system such that when the cooling demand is low, we can save on pumping power while increasing the pumping power only when demand is high.

The work in [58] investigated a micro-channel based DTM scheme that could provide sufficient cooling to the 3D-IC using minimal amount of cooling energy. In this DTM scheme, assuming the micro-channel structure is fixed, the DTM scheme dynamically controls the pressure drop across the micro-channels (basically the fluid flow rate) based on the runtime cooling demand. Here we will explain the micro-channel based DTM scheme proposed in [58] in detail.

The temperature profiles on chip is a strong function of the power dissipated, while the power dissipation depends on the applications which change at runtime. In order to track the runtime thermal and power state, thermal sensors are placed at various chip locations. The DTM scheme proposed in [58] keeps track of power profiles at runtime using the information achieved by thermal sensors and adaptive Kalman filter-based estimation approach (proposed in [59]), and then decides the micro-channel fluid flow rate based on it.

To estimate the power profile, Zhang and Srivastava [59] assumes there are $M$ different power states (power profiles), each of which essentially represents a certain class of applications. The Kalman filter holds a belief of what the current power profile is and predict the temperature profile based on this belief. Meanwhile, the thermal sensors keep measuring the temperature. The power estimation method in [59] iteratively compares the temperature predicted by Kalman filter and sensor observations. If the error between them are close to zero, this indicates that the belief of current power state is correct. Otherwise, the belief might be wrong, which means the power state has changed. Once the change in power state is detected, it tries to decide the new power state, which is the one most likely to result in the current sensor reading. We are not going into details of this power estimation approach, since this is not the focus of this article. Interested readers are referred to [59] for the details of this adaptive power estimation approach.

Once the power profile is obtained, the micro-channel based DTM approach in [58] will select the best pressure drop which provides enough cooling for this power profile using minimum pumping power. Hence the micro-channel based DTM problem is formally stated as follows.

*Given:* A 3D-IC design, its power distribution is a function of the architecture and application. Assuming the power profiles are given and the micro-channel structure is also fixed, the DTM scheme would like to find the pressure drop for each power profile such that the temperature across the chip is within acceptable limits while minimizing pumping power:

$$\min \quad Q_{\mathrm{pump}}(\Delta P) = \left( \frac{Nb\, d\, D_h^2}{2\gamma\mu L} \right) \Delta P^2 \quad \Leftrightarrow \quad \min \Delta P.$$

$$\text{s.t.} \quad H(\Delta P) \cdot T = Q, \tag{25}$$

$$T \leqslant T_{\max},$$

$$\Delta P_{\min} \leqslant \Delta P \leqslant \Delta P_{\max},$$

The objective minimizes the pumping power used by micro-channels. Here regular straight micro-channels are used, hence the pumping power can be calculated using Eqns (17) and (18). The first constraint indicates the resistive thermal model, where $Q$ is a 3D-IC power profile and $T$ is the corresponding thermal profile, and $H$ is the thermal conductivity matrix which depends on pressure drop $\Delta P$. The second constraint indicates that the peak temperature should not exceed the thermal constraint $T_{\max}$. The last constraint gives the feasible range of pressure drop.

This optimization problem is difficult to solve directly because of the complexity of thermal model and the impact of micro-channel on temperature. Therefore Shi and Srivastava [58] uses a linear search based heuristic to find the best pressure drop. Assume the micro-channel structure is already decided, therefore the pumping power $Q_{\mathrm{pump}}$ is only a function of pressure drop $\Delta P$ in this problem. From the objective in Eqn (25), pumping power is a quadratic function of pressure drop, which means the pumping power increases monotonically as pressure drop increases. Hence minimizing pressure drop basically minimizes pumping power, and the problem is simplified to finding the minimum pressure drop that provides enough cooling.

Pressure drop $\Delta P$ influences heat resistance $R_{\mathrm{heat}}$, thereby changing the cooling performance. From Eqns (13) and (18), we can find that given the micro-channel design, increase in pressure drop $\Delta P$ results in increased fluid velocity $V$ and flow rate $F$, while higher flow rate results in smaller heat resistance $R_{\mathrm{heat}}$ and hence better cooling performance.

In summary, a larger $\Delta P$ would result in better cooling at the cost of higher pumping power. Hence cooling effectiveness is a monotonic function of $\Delta P$. Therefore the linear search approach can find the best pressure drop. Specifically, this is done by starting from the minimum pressure drop $\Delta P = \Delta P_{min}$ and increasing it step by step until thermal constraint is satisfied. Due to the monotonic nature of the impact of pressure drop on micro-channel cooling effectiveness, this linear search approach can result in the optimal selection of pressure drop for a given micro-channel allocation.

## 7. CONCLUSION

This chapter investigated several aspects of micro-fluidic cooling in stacked 3D-ICs. We studied the structure of 3D-IC with micro-channels, and its thermal/hydrodynamic modeling approaches. We also discussed the design considerations and challenges of micro-channel heat sinks in 3D-IC, and summarized existing micro-channel infrastructures. In the last part, we introduced a DTM scheme using micro-fluidic cooling that can control the runtime chip temperature by dynamically tuning the fluid flow rate through micro-channels, such that sufficient cooling can be provided using minimal cooling power.

### ACKNOWLEDGMENT

We would like to thank NSF grants CCF 0937865 and CCF 0917057 for supporting part of this research.

### REFERENCES

[1] M.S. Bakir, C. King, et al., 3D heterogenous integrated systems: liquid cooling, power delivery, and implementation, in: IEEE Custom Intergrated Circuits Conference, 2008, pp. 663–670.

[2] D.B. Tuckerman, R.F.W. Pease, High-performance heat sinking for VLSI, IEEE Electron Device Letters (1981) 126–129.

[3] Y.J. Kim, Y.K. Joshi, et al., Thermal characterization of interlayer microfluidic cooling of three dimensional integrated circuits with nonuniform heat flux, ASME Transactions Journel of Heat Transfer (2010).

[4] J.-M. Koo, S. Im, L. Jiang, K.E. Goodson, Integrated microchannel cooling for three-dimensional electronic circuit architectures, ASME Transactions Journel of Heat Transfer (2005) 49–58.

[5] R.W. Knight, D.J. Hall, et al., Heat sink optimization with application to microchannels, IEEE Transaction on Components, Hybrids, and Manufacturing Technology (1992) 832–842.

[6] F.M. White, Fluid Mechanics, McGraw-Hill Book Company, 1986.

[7] W. Qu, I. Mudawar, S.-Y. Lee, S.T. Wereley, Experimental and computational investigation of flow development and pressure drop in a rectangular micro-channel, Journal of Electronic Packaging (2006).

[8]  R. Walchli, T. Brunschwiler, B. Michel, D. Poulikakos, Combined local microchannel-scale CFD modeling and global chip scale network modeling for electronics cooling design, International Journal of Heat and Mass Transfer (2010).

[9]  D. Kim, S.J. Kim, A. Ortega, Compact modeling of fluid flow and heat transfer in pin fin heat sinks, Journal of Electronic Packaging (2004).

[10]  A. Sridhar, A. Vincenzi, M. Ruggiero, T. Brunschwiler, D. Atienza, 3D-ICE: fast compact transient thermal modeling for 3D ICS with inter-tier liquid cooling, in: IEEE/ACM International Conference on Computer Aided Design (ICCAD'10), 2010.

[11]  B.R. Munson, D.F. Young, T.H. Okiishi, W.W. Huebsch, Fundamentals of Fluid mechanics, Wiley, 2008.

[12]  S. Choi, R. Barron, R. Warrington, Fluid flow and heat transfer in micro tubes, in: Micromechanical Sensors, Actuators and Systems, ASME DSC, 1991, pp. 123–128.

[13]  T.M. Adams, S.I. Abdel-Khalik, S.M. Jeter, Z.H. Qureshi, An experimental investigation of single-phase forced convection in microchannels, International Journal of Heat and Mass Transfer (1998) 851–857.

[14]  P.Y. Wu, W. Little, Measuring of the heat transfer characteristics of gas flow in fine channel heat exchangers for micro miniature refrigerators, Cryogenics (1994).

[15]  D. Yu, R. Warrington, R. Barron, T. Ameen, An experimental and theoretical investigation of fluid flow and heat transfer in microtubes, in: Proceedings of the ASME/JSME Thermal Engineering Conference, 1995, pp. 523–530.

[16]  J.R. Thome, Engineering Data Book iii, Wolverine Tube, 2004.

[17]  I.H.M. Dang, R. Muwanga, Adiabatic two phase flow distribution and visualization in scaled microchannel heat sinks, Experiments in Fluids, (2007).

[18]  B. Agostini, J.R. Thome, M. Fabbri, B. Michel, High heat flux two-phase cooling in silicon multimicrochannels, IEEE Transactions on Components and Packaging Technologies 31 (2008).

[19]  S. Senn, D. Poulikakos, Laminar mixing, heat transfer and pressure drop in tree-like microchannel nets and their application for thermal management in polymer electrolyte fuel cells, Journal of Power Sources 130 (2004) 178–191.

[20]  Y.S. Muzychka, M.M. Yovanovich, Modelling friction factors in non-circular ducts for developing laminar flow, in: Second AIAA Theoretical Fluid Mechanics Meeting, 1998.

[21]  A. Bejan, Convection Heat Transfer, Wiley, New York, 1994.

[22]  R.J. Phillips, Micro-channel heat sinks, Advances in Thermal Modeling of Electronic Components and Systems, vol. 2, ASME, New York, 1990.

[23]  C. King, D. Sekar, M. Bakir, B. Dang, J. Pikarsky, J. Meindl, 3D stacking of chips with electrical and microfluidic I/O interconnects, in: 58th Electronic Components and Technology Conference (ECTC'08), 2008, pp. 1–7.

[24]  Y.-J. Lee, S.K. Lim, Co-optimization of signal, power, and thermal distribution networks for 3D IC, in: Electrical Design of Advanced Packaging and Systems Symposium, 2008, pp. 163–155.

[25]  K. Puttaswamy, G.H. Loh, Thermal analysis of a 3d die-stacked high-performance microprocessor, in: Proceedings of the 16th ACM Great Lakes Symposium on VLSI (GLSVLSI'06 ), 2006.

[26]  Y. Zhang, A. Srivastava, Accurate temperature estimation using noisy thermal sensors for Gaussian and non-Gaussian cases, IEEE Transactions on Very Large Scale Integration (VLSI) Systems 19 (2011) 1617–1626.

[27]  G. Hamerly, E. Perelman, J. Lau, B. Calder, Simpoint 3.0: Faster and more flexible program analysis, Journal of Instruction Level Parallelism (2005).

[28]  M.B. Healy, S.K. Lim, Power delivery system architecture for many-tier 3d systems, in: Electronic Components and Technology Conference, 2010, pp. 1682–1688.

[29]  M. Pathak, Y.-J. Lee, T. Moon, S.K. Lim, Through-silicon-via management during 3d physical design: when to add and how many? in: IEEE/ACM International Conference on Computer-Aided Design (ICCAD'10), 2010, pp. 387–394.

[30] J.-S. Yang, K. Athikulwongse, Y.-J. Lee, S.K. Lim, D.Z. Pan, TSV stress aware timing analysis with applications to 3D-IC layout optimization, in: Proceedings of the 47th Design Automation Conference (DAC'10), 2010.

[31] K. Athikulwongse, A. Chakraborty, J.-S. Yang, D. Pan, S. K. Lim, Stress-driven 3D-IC placement with tsv keep-out zone and regularity study, in: IEEE/ACM International Conference on Computer Aided Design (ICCAD'10), 2010.

[32] H. Irie, K. Kita, K. Kyuno, A. Toriumi, In-plane mobility anisotropy and universality under uni-axial strains in n- and p-MOS inversion layers on (1 0 0), (1 1 0), and (1 1 1) si, in: IEEE International Electron Devices Meeting, 2004, pp. 225–228.

[33] C.S. Smith, Piezoresistance effect in germanium and silicon, Physical Review 94 (1954) 42–49.

[34] <http://www.ansys.com/Products/Simulation+Technology/Fluid+Dynamics/ANSYS+CFX>.

[35] A. Sridhar, A. Vincenzi, M. Ruggiero, T. Brunschwiler, D. Atienza, Compact transient thermal model for 3D ICS with liquid cooling via enhanced heat transfer cavity geometries, in: Proceedings of the 16th International Workshop on Thermal Investigations of ICs and Systems (THERMINIC'10), 2010.

[36] K. Skadron, M.R. Stan, K. Sankaranarayanan, W. Huang, S. Velusamy, D. Tarjan, Temperature-aware microarchitecture: modeling and implementation, ACM Transactions on Architecture and Code Optimization 1 (2004) 94–125, 3.

[37] B. Shi, A. Srivastava, P. Wang, Non-uniform micro-channel design for stacked 3D-ICS, in: Design Automation Conference (DAC'11), 2011.

[38] S. Kandlikar, S. Garimella, et al., Heat Transfer and Fluid flow in Minichannels and Microchannels, Elsevier, 2005.

[39] R.K. Shah, A.L. London, Laminar Flow Forced Convection in Ducts: A Source Book for Compact Heat Exchanger Analytical Data, Academic Press, 1978.

[40] A. Husain, K.-Y. Kim, Shape optimization of micro-channel heat sink for microelectronic cooling, IEEE Transactions on Components and Packaging Technologies (2008) 322–330.

[41] X. Wei, Y. Joshi, Optimization study of stacked micro-channel heat sinks for microelectronic cooling, IEEE Transactions on Components and Packaging Technologies (2003) 55–61.

[42] J.P. McHale, S.V. Garimella, Heat transfer in trapezoidal microchannels of various aspect ratios, International Journal of Heat and Mass Transfer (2009) 365–375.

[43] W. Qu, M. Mala, D. Li, Pressure-driven water flows in trapezoidal silicon microchannels, International Journal of Heat and Mass Transfer (2000) 353–364.

[44] L. Jiang, J.-M. Koo, et al., Cross-linked microchannels for VLSI hotspot cooling, in: ASME 2002 International Mechanical Engineering Congress and Exposition, 2002.

[45] C. Marques, K.W. Kelly, Fabrication and performance of a pin fin micro heat exchanger, Journal of Heat Transfer (2004) 434–444.

[46] A. Kosar, C. Mishra, Y. Peles, Laminar flow across a bank of low aspect ration micro pin fins, Journal of Fluids Engineering (2005) 419–430.

[47] Y. Peles, A. Kosar, C. Mishra, C.-J. Kuo, B. Schneider, Forced convective heat transfer across a pin fin micro heaet sink, International Journal of Heat and Mass Transfer (2005) 3615–3627.

[48] Y. Chen, P. Cheng, Heat transfer and pressure drop in fractal tree-like microchannel nets, International Journal of Heat and Mass Transfer 45 (2002) 2643–2648.

[49] L. Ghodoossi, Thermal and hydrodynamic analysis of a fractal microchannel network, Energy Conversion and Management (2005) 71–788.

[50] X.-Q. Wang, A.S. Mujumdar, C. Yap, Thermal characteristics of tree-shaped microchannel nets for cooling of a rectangular heat sink, International Journal of Thermal Sciences 45 (2006) 1103–1112.

[51] T. Brunschwiler, B. Michel, H. Rothuizen, U. Kloter, B. Wunderle, H. Reichl, Hotspot-optimized interlayer cooling in vertically integrated packages, in: Proceedings of the Materials Research Society (MRS) Fall Meeting, 2008.

[52] T. Brunschwiler, S. Paredes, U. Drechsler, B. Michel, W. Cesar, Y. Leblebici, B. Wunderle, H. Reichl, Heat-removal performance scaling of interlayer cooled chip stacks, in: 12th IEEE Intersociety Conference on Thermal and Thermomechanical Phenomena in Electronic Systems (ITherm), 2010, pp. 1–12.

[53] T. Brunschwiler, B. Michel, H. Rothuizen, U. Kloter, B. Wunderle, H. Oppermann, H. Reichl, Interlayer cooling potential in vertically integrated packages, Microsystem Technologies 15 (2009) 57–74.

[54] B. Shi, A. Srivastava, TSV-constrained micro-channel infrastructure design for cooling stacked 3D-ICS, in: International Symposium on Physical Design (ISPD'12), 2012.

[55] A.K. Coskun, J.L. Ayala, D. Atienzaz, T.S. Rosing, Modeling and dynamic management of 3D multicore systems with liquid cooling, in: 17th Annual IFIP/IEEE International Conference on Very Large Scale Integration, 2009, pp. 60–65.

[56] A.K. Coskun, D. Atienza, T.S. Rosing, et al., Energy-efficient variable-flow liquid cooling in 3D stacked architectures, in: Conference on Design, Automation and Test in Europe (DATE'10), 2010, pp. 111–116.

[57] H. Qian, X. Huang, H. Yu, C.H. Chang, Cyber-physical thermal management of 3d multi-core cache professor system with microfluidic cooling, Journal of Low Power Electronics, 2011.

[58] B. Shi, A. Srivastava, Cooling of 3D-IC using non-uniform micro-channels and sensor based dynamic thermal management, in: 49th Annual Allerton Conference on Communication, Control, and Computing, 2011, pp. 1400–1407.

[59] Y. Zhang, A. Srivastava, Adaptive and autonomous thermal tracking for high performance computing systems, in: Design Automation Conference (DAC'10), 2010.

## ABOUT THE AUTHORS

**Bing Shi** received the B.S. degree from Zhejiang University, Hangzhou, China, in 2005, and the M.S. degree in electrical and computer engineering from the University of Maryland, College Park, in 2012, where she is currently pursuing the Ph.D. degree in computer engineering. Her current research interests include thermal and power management for high-performance computing systems, 3D-IC thermal-related issues, and physical design.

Ms. Shi was the recipient of the ISVLSI 2012 Best Paper Award, the Distinguished Graduate Fellowship, and Summer Research Fellowship from the University of Maryland.

**Ankur Srivastava** received his B.Tech in Electrical Engineering from Indian Institute of Technology Delhi in 1998, M.S. in ECE from Northwestern University in 2000 and PhD in Computer Science from UCLA in 2002. He is currently an Associate Professor in the ECE department with joint appointment with the Institute for Systems Research at University of Maryland, College Park. His primary research interest lies in the field of high performance, low power electronic systems and applications such as computer vision, data and storage centers and sensor networks, three-dimensional circuits, as well as hardware security.

Dr. Srivastava is the associate editor of IEEE Transactions on VLSI and INTEGRATION: VLSI Journal. He has served on several NSF panels and was in the technical program committee of several conferences such as ICCAD, DAC, ISPD, ICCD, GLSVLSI. His research and teaching endeavors have received several awards including the Outstanding PhD Dissertation Award from the CS Department of UCLA, Best Paper Award from ISPD 2007, George Corcoran Memorial Teaching Award from the ECE Department of University of Maryland, Best Paper Nomination in ICCAD 2003, ACM-SIGDA Outstanding Dissertation Nomination in 2003.

# CHAPTER THREE

# Sustainable DVFS-enabled Multi-Core Architectures with on-chip Wireless Links

Jacob Murray, Teng Lu, Partha Pande, and Behrooz Shirazi
School of Electrical Engineering and Computer Science, Washington State University, Pullman, WA 99164 USA

## Contents

## Abstract

Wireless Network-on-Chip (WiNoC) has emerged as an enabling technology to design low power and high bandwidth massive multi-core chips. The performance advantages mainly stem from using the wireless links as long-range shortcuts between far apart cores. This performance gain can be enhanced further if the characteristics of the wireline links and the processing cores of the WiNoC are optimized according to the traffic patterns and workloads. This chapter demonstrates that by incorporating both

*Advances in Computers*, Volume 88
ISSN 0065-2458, http://dx.doi.org/10.1016/B978-0-12-407725-6.00003-4

processor- and network-level dynamic voltage and frequency scaling (DVFS) in a WiNoC, the power and thermal profiles can be enhanced without a significant impact on the overall execution time.

# 1. INTRODUCTION

The continuing progress and integration levels in silicon technologies make possible complete end-user systems on a single chip. An important performance limitation of massive multi-core chips designed with traditional network fabrics arises from planar metal interconnect-based multi-hop links, where the data transfer between two distant blocks can cause high latency and power consumption. Increased power consumption will give rise to higher temperature, which in turn can decrease chip reliability and performance and increase cooling costs. Different approaches for creating low-latency, long-range communication channels like optical interconnects, on–chip transmission lines, and wireless interconnects have been explored. These emerging interconnect technologies can enable the design of so called small-world on-chip network architectures, where closely spaced cores will communicate through traditional metal wires, but long-distance communications will be predominantly achieved through high-performance specialized links [1].

One possible innovative and novel approach is to replace multi-hop wireline paths in a Network-on-Chip (NoC) by high-bandwidth single-hop long-range wireless links [2, 3]. The on–chip wireless links facilitate design of a small-world NoC by enabling one-hop data transfers between distant nodes. In addition to reducing interconnect delay and eliminating multi-hop long-distance wireline communication, it reduces the energy dissipation as well. However, the overall energy dissipation of the wireless NoC is still dominated by wireline links. The link utilization varies depending on the on-chip communication patterns. Tuning the link bandwidth accurately to follow the traffic requirements opens up possibilities of significant power savings. Dynamic voltage and frequency scaling (DVFS) is a well-known technique that enables adjusting the bandwidth of the interconnects by suitably varying their voltage and frequency levels. Consequently, this will enable power savings and lowering of temperature hotspots in specific regions of the chip. In this chapter the aim is to show how a small-world NoC architecture with long-range wireless links and DVFS-enabled wireline interconnects lowers the energy dissipation of a multi-core chip, and consequently helps to improve the thermal profile. This chapter looks at the current limitations of NoC with respect to latency, power consumption, and thermal issues. This is first covered by explaining existing works on this topic. Then, a WiNoC architecture is proposed to address network level latency and power problems. The network is then further optimized for power and temperature constraints using DVFS. After addressing the network issues, the processing cores are also addressed by implementing DVFS among them as well. From thermal results, the processing elements are further optimized by placing the hottest cores along the chip edge in order to further improve the thermal profile of the chip.

## 2. RELATED WORK

The limitations and design challenges associated with existing NoC architectures are elaborated in [4]. This chapter highlights interactions among various open research problems of the NoC paradigm. Conventional NoCs use multi-hop packet switched communication. At each hop the data packet goes through a complex router/switch, which contributes considerable power consumption and, throughput and latency overhead. To improve performance, the concept of express virtual channels is introduced in [5]. By using virtual express lanes to connect distant cores in the network, it is possible to avoid the router overhead at intermediate nodes, and thereby improve NoC performance in terms of power, latency, and throughput. Performance is further improved by incorporating ultra low-latency, multi-drop on-chip global lines (G-lines) for flow control signals. NoCs have been shown to perform better by insertion of long-range wired links following principles of small-world graphs. Despite significant performance gains, the above schemes still require laying out long wires across the chip and hence performance improvements beyond a certain limit cannot be achieved.

Designs of small-world based hierarchical wireless NoC (WiNoC) architectures were introduced in [2, 3]. Recently, the design of a WiNoC based on CMOS ultra wideband (UWB) technology was proposed [6]. In [7], the feasibility of designing miniature antennas and simple transceivers that operate in the sub-terahertz frequency range for on-chip wireless communication has been demonstrated. In [8] a combination of Time and Frequency Division Multiplexing is used to transfer data over inter-router wireless express channels.

All the aforementioned works demonstrate advantages of wireless NoC architectures in terms of latency and energy dissipation. But, none of those address the correlation between the energy dissipation and temperature profile of a wireless NoC. In this chapter, the aim is to bridge that gap by quantifying how a small-world based wireless NoC architecture reduces the heat dissipation of a multi-core chip in addition to improving the latency and energy dissipation characteristics.

Most of the existing works related to the design of wireless NoC demonstrate its advantages in terms of latency and energy dissipation provided by the wireless channels only. The main emphasis always has been on the characteristics of the wireless links. However, the overall energy dissipation of the wireless NoC can be improved even further if the characteristics of the wireline links are optimized as well depending on their utilization requirements based on the traffic patterns. DVFS is known to be an efficient technique to reduce energy dissipation of interconnection networks.

DVFS is a popular methodology to optimize the power dissipation of electronic systems without significantly compromising overall system performance [9]. DVFS has also been proven to be an effective approach in reducing energy consumption of single processors [10]. DVFS can be applied to multi-core processors either to all cores or to individual cores independently [11]. There are many commercially available voltage-scalable processors, including Intel's Xscale, Transmeta's Cruso, and AMD's mobile processors [12]. Multi-core chips implemented with multiple Voltage Frequency Island (VFI) design style is another promising alternative. It is shown to be effective in reducing on-chip power dissipation [13, 14]. Designing appropriate DVFS control algorithms for VFI systems has been addressed by various research groups [15]. Some researchers have also recently discussed the practical aspects of implementing DVFS control on a chip, such as tradeoffs between on-chip versus off-chip DC-DC converters [11] and the number of allowed discrete voltage levels [16] and centralized versus distributed control techniques [17]. DVFS also helps to improve the thermal profile of multi-core processors. Thermal-aware techniques are principally related to power-aware design methodologies [18]. It is shown that distributed DVFS provides considerable performance improvement under thermal duress [18]. When independent per-core DVFS is not available, then thread migration performs well. Temperature-aware scheduling techniques for multi-core processors are also investigated by various research groups [18].

Most of these works principally addressed power and thermal management strategies for the processing cores only. Networks consume a significant part of the chip's power budget; greatly affecting overall chip temperature. However, there is little research on how they contribute to thermal problems [19]. Thermal Herd, a distributed runtime scheme for thermal management that lets routers collaboratively regulate the network temperature profile and work to avert thermal emergencies while minimizing performance impact was proposed in [19]. In [12], for the first time, the problem of simultaneous dynamic voltage scaling of processors and communication links for real-time distributed systems has been addressed. Intel's recent multicore-based single chip cloud computers (SCC) incorporate DVFS both in the core and the network levels. But, all of the aforementioned works principally considered standard multi-hop interconnection networks for the multi-core chips, the limitations of what are well known. The principal emphasis has been on the design of DVFS algorithms, voltage regulators along with scheduling and flow control mechanisms. It has been shown how efficient on-chip network design in conjunction with novel interconnect technologies can open a new direction in design of low power and high bandwidth multi-core chips [20]. It is already demonstrated that small-world network architectures with long-range wireless shortcuts can significantly improve the energy consumption and achievable data rate of massive multi-core computing platforms [21]. In this chapter the aim is to show how a small-world wireless NoC architecture with DVFS-enabled wireline links improves the energy dissipation and thermal profile of a multi-core chip.

## 3. PROPOSED ARCHITECTURE

Traditionally, a mesh is the most popular NoC architecture due to its simplicity and the regularity of grid structure. However, one of the most important limitations of this architecture is the multi-hop communications between far apart cores, which gives rise to significant latency and energy overheads. To alleviate these shortcomings, long-range and single-hop wireless links are inserted as shortcuts on top of a mesh. It is shown that insertion of long-range wireless shortcuts in a conventional wireline NoC has the potential for bringing significant improvements in performance and energy dissipation [2, 3]. These works showed that over a traditional mesh, a small-world network can improve throughput and energy dissipation both by orders of magnitude. Inserting the long-range links in a conventional wireline mesh reduces the average hop count, and increases the overall connectivity of the NoC. The architecture is referred to as the wireless mesh (WiMesh). In this WiMesh architecture, by careful placement of wireless links depending on distance and traffic patterns savings in latency, energy, and heat dissipation are enabled. In addition, by implementing per-link history-based DVFS on energy-inefficient wireline links, the DVFS-enabled WiMesh (D-WiMesh) further improves upon savings in energy and heat dissipation by dynamically reducing voltage and frequency depending on the bandwidth requirements of the application. However, it sacrifices the latency improvements introduced by the wireless shortcuts due to mispredicted traffic requirements and switching penalty to some extent. All of these design trade-offs and performance evaluations are presented in the following subsections.

## 3.1 Physical Layer Design

Suitable on-chip antennas are necessary to establish wireless links for WiNoCs. In [22] the authors demonstrated the performance of silicon integrated on-chip antennas for intra- and inter-chip communication. They have primarily used metal zig-zag antennas operating in the range of tens of GHz. This particular antenna was used in the design of a wireless NoC [3] mentioned earlier in Section 2. The aforementioned antennas principally operate in the millimeter wave (tens of GHz) range and consequently their sizes are on the order of a few millimeters.

If the transmission frequencies can be increased to THz/optical range then the corresponding antenna sizes decrease, occupying much less chip real estate. Characteristics of metal antennas operating in the optical and near-infrared region of the spectrum of up to 750 THz have been studied [23]. Antenna characteristics of carbon nanotubes (CNTs) in the THz/optical frequency range have also been investigated both theoretically and experimentally [24, 25]. Bundles of CNTs are predicted to enhance performance of antenna modules by up to 40 dB in radiation efficiency and provide excellent directional properties in far-field patterns [26]. Moreover these antennas can achieve a bandwidth of around 500 GHz, whereas the antennas operating in the millimeter wave range achieve bandwidths of tens of GHz [26]. Thus, antennas operating in the THz/optical frequency range can support much higher data rates. CNTs have numerous characteristics that make them suitable as on-chip antenna elements for optical frequencies. Given wavelengths of hundreds of nanometers to several micrometers, there is a need for virtually one-dimensional antenna structures for efficient transmission and reception. With diameters of a few nanometers and any length up to a few millimeters possible, CNTs are the perfect candidate. Such thin structures are almost impossible to achieve with traditional microfabrication techniques for metals. Virtually defect-free CNT structures do not suffer from power loss due to surface roughness and edge imperfections found in traditional metallic antennas. In CNTs, ballistic electron transport leads to quantum conductance, resulting in reduced resistive loss, which allows extremely high current densities in CNTs, namely 4–5 orders of magnitude higher than copper. This enables high transmitted powers from nanotube antennas, which is crucial for long-range communications. By shining an external laser source on the CNT, radiation characteristics of multi-walled carbon nanotube (MWCNT) antennas are observed to be in excellent quantitative agreement with traditional radio antenna theory [24], although at much higher frequencies of hundreds of THz. Using various lengths of the antenna elements corresponding to different multiples of the wavelengths of the external lasers, scattering and radiation patterns are shown to be improved [24]. Such nanotube antennas are good candidates for establishing on-chip wireless communications links and are henceforth considered in this work.

Chemical vapor deposition (CVD) is the traditional method for growing nanotubes in specific on-chip locations by using lithographically patterned catalyst islands. The application of an electric field during growth or the direction of gas flow during CVD can help align nanotubes. However, the high-temperature CVD could potentially damage some of the pre-existing CMOS layers. To alleviate this, localized heaters in the CMOS fabrication process to enable localized CVD of nanotubes without exposing the entire chip to high temperatures are used [27]. The wireless nodes need to be equipped with transmitting and receiving antennas, which will be excited using external laser sources. As mentioned in [28], the laser sources can be located off-chip or bonded to the silicon die. Hence their power dissipation does not contribute to the chip power density. The requirements of using external sources to excite the antennas can be eliminated if the electroluminescence phenomenon from a CNT is utilized to design linearly polarized dipole radiation sources [29]. But further investigation is necessary to establish such devices as successful transceivers for on-chip wireless communication.

To achieve line of sight communication between using CNT antennas at optical frequencies, the chip packaging material has to be elevated from the substrate surface to create a vacuum for transmission of the high frequency electromagnetic (EM) waves. Techniques for creating such vacuum packaging are already utilized for Micro-electro-mechanical Systems (MEMS) applications [30], and can be adopted to make creation of line of sight communication between CNT antennas viable. In classical antenna theory it is known that the received power degrades inversely with the fourth power of the separation between source and destination due to ground reflections beyond a certain distance. This threshold separation, $r_0$ between source and destination antennas assuming a perfectly reflecting surface, is given by (1):

$$r_0 = \frac{2\pi H^2}{\lambda}. \tag{1}$$

Here $H$ is the height of the antenna above the reflecting surface and $\lambda$ is the wavelength of the carrier. Thus, if the antenna elements are at a distance of $H$ from the reflective surfaces like the packaging walls and the top of the die substrate, the received power degrades inversely with the square of the distance. Thus $H$ can be adjusted to make the maximum possible separation smaller than the threshold separation $r_0$ for a particular frequency of radiation used. Considering the optical frequency ranges of CNT antennas, depending on the separation between the source and destination pairs in a single chip, the required elevation is a few tens of microns only.

## 3.2 Wireless Link Placement

The placement of the wireless links between a particular pair of source and destination routers is important as this is responsible for establishing high-speed, low-energy interconnects on the network, which will eventually result in performance gains. Realistically, there may be non-uniform traffic distributions within the network. Therefore, the links are placed probabilistically; i.e. between each pair of source and destination routers, $i$ and $j$, respectively, where the probability $P_{ij}$ of having a wireless link is proportional to the distance measured in number of hops between routers, $i$ and $j$, $h_{ij}$, and the amount of communication between routers, $i$ and $j$, $f_{ij}$, as shown in (2):

$$P_{ij} = \frac{h_{ij} f_{ij}}{\sum_{i,j} h_{ij} f_{ij}}. \tag{2}$$

In (2), $f_{ij}$ is the frequency of communication between the $i$th source and $j$th destination. This frequency is expressed as the percentage of traffic generated from $i$ that is addressed to $j$. This frequency distribution is based on the particular application mapped to the overall NoC and is hence set prior to wireless link insertion. Therefore, the a priori knowledge of the traffic pattern is used to optimize the WiNoC. This optimization approach establishes a correlation between traffic distribution across the NoC and network configuration as in [31]. The probabilities are normalized such that their sum is equal to one, where $\mu$ is the optimization metric (3):

$$\mu = \sum_{i,j} h_{ij} f_{ij}. \tag{3}$$

Equal weight is attached to distance as well as frequency of communication in the metric. Such a distribution is chosen because in the presence of a wireless link, the distance between the pair becomes a single hop and hence it reduces the original distance between the communicating routers in the network. Thus, highly communicating routers have a higher need for a wireless link.

Depending on the number of available wireless links, they are inserted between randomly chosen pairs of routers, which are chosen following the probability distribution mentioned above. Once the network is initialized, an optimization by means of simulated annealing (SA) heuristics is performed. The optimization step is necessary as the random initialization might not produce the optimal network topology. SA offers a simple, well established, and scalable approach for the optimization process as opposed to a brute-force search. If there are $N$ routers in the network and $n$ wireless links to distribute, the size of the search space $S$ is given by (4):

$$|S| = \left( \begin{array}{c} \binom{N}{2} - N \\ n \end{array} \right).$$

$$(4)$$

Thus, with increasing $N$, it becomes increasingly difficult to find the best solution by exhaustive search. In order to perform SA, a metric has been established, which is closely related to the connectivity of the network. The metric to be optimized (3) is $\mu$. To compute this metric, the shortest distances between all router pairs are computed following the routing strategy outlined later in this section. With each iteration of the SA process, a new network is created by randomly rewiring a wireless link in the current network. The metric for this new network is calculated and compared to the metric of the current network. The new network is always chosen as the current optimal solution if the metric is lower. However, even if the metric is higher, the new network is chosen probabilistically. This reduces the probability of getting stuck in a local optimum, which could happen if the SA process were to never choose a worse solution. The exponential probability shown in (5) is used to determine whether or not a worse solution is chosen as the current optimal:

$$P(\mu,\mu',T) = e^{\frac{\mu-\mu'}{T}}.$$

$$(5)$$

The optimization metrics (3) for the current and new networks are $\mu$ and $\mu'$, respectively. $T$ is a temperature parameter, which decreases with the number of optimization iterations according to an annealing schedule. In this work, Cauchy scheduling is used, where the temperature varies inversely with the number of iterations [32]. The algorithm used to optimize the network is shown in Fig. 1, which explains the flow of the SA algorithm from above. Given the probabilities of traffic distributions in the network, the metric for optimization uses the known traffic distributions as a method for wireless link insertion. For each application specific traffic pattern, the wireless link insertion process finds a corresponding optimum network configuration. Due to this inherent correlation between the optimization of the network configuration and the traffic pattern, the gains in throughput with non-uniform traffic are larger than that with uniform traffic. An important component in the design of the WiNoCs is the on-chip antenna for the wireless links.

By using multi-band laser sources to excite CNT antennas, different frequency channels can be assigned to pairs of communicating source and destination nodes. This will require using antenna elements tuned to different frequencies for each pair, thus creating a form of frequency division multiplexing (FDM) creating dedicated channels between a source and destination pair. This is possible by using CNTs of different lengths, which are multiples of the wavelengths of the respective carrier frequencies. High directional gains of these antennas, demonstrated in [24], aid in creating directed channels between source and destination pairs. In [33], 24 continuous wave laser sources of different frequencies are used. Thus, these 24 different frequencies can be assigned to multiple wireless links in the WiMesh in such a way that a single frequency channel is used only once to avoid signal interference on the same frequencies. This enables concurrent use of multi-band channels over the chip. The placement of the links is dependent upon three main parameters, the number of cores, the number of long-range links to be placed, and the traffic distribution. The aim of the wireless link placement is to minimize the hop count of the network. As discussed above, the average hop count is optimized, weighted by the probability of traffic interactions among the cores. In this way equal importance is attached to both inter-core distance and frequency of communication. A single hop in this work is defined as the path length between a source and destination pair that can be traversed in one clock cycle. Wireless link placement is crucial for optimum performance gain as it establishes high-speed, low-energy interconnects on the network. Placement of wireless links in a NoC, by using the simulated annealing (SA)-based methodology converges to the optimal configuration much faster than the exhaustive search technique. Hence, a

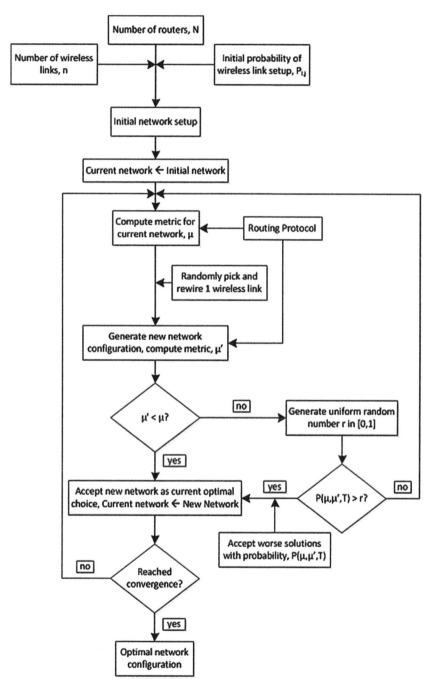

**Fig. 1.** Network optimization algorithm.

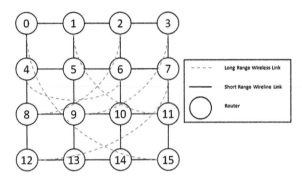

**Fig. 2.** Sample wireless link placement, 16-core FFT.

SA-based optimization technique is adopted for placement of the wireless links to get maximum benefits of using the wireless shortcuts. Care is taken to ensure that introduction of the wireless shortcuts does not give rise to deadlocks in data exchange. The routing adopted here is a combination of dimension order (*X-Y*) routing for the nodes without wireless links and South-East routing algorithm for the nodes with wireless shortcuts. This routing algorithm is proven to be deadlock free in [1]. Figure 2 shows a sample link placement of a 16-core WiMesh network with Fast Fourier Transform (FFT)-based traffic after running the SA algorithm. As can be seen, by running the SA algorithm until the optimization metric from (3) was minimized, wireless links are inserted from router 0 to router 15, 1 to 11, 2 to 8, 3 to 9, 4 to 6, and 7 to 12, where the number of routers was 16, and the number of wireless links was 6.

## 3.3 Dynamic Voltage and Frequency Scaling

It has already been shown that by introducing wireless links as long-range shortcuts onto a conventional wireline NoC, savings in latency and energy is obtained [2, 3]. By reducing the hop count between largely separated communicating cores, wireless shortcuts have been shown to attract a significant amount of the overall traffic within the network [2]. The amount of traffic detoured is substantial and the low power wireless links enable energy savings. However, the energy dissipation within the network is still dominated by the flits traversing the wireline links. One popular method to reduce energy consumption of these wireline links is to incorporate DVFS. A method for history-based link level DVFS was proposed in [34]. In this scheme, every NoC router predicts future traffic patterns based on what was seen in the past. The metric to determine whether DVFS should be performed is link utilization. The short-term link utilizations characterized by (6):

$$U_{\text{Short}} = \frac{1}{H} \sum_{i=1}^{H} f_i,$$  (6)

where $H$ is the history window, and $f_i$ is 1 if a flit traversed the link on the $i$th cycle of the history window and a 0 otherwise. The predicted future link utilization, $U_{\text{Predicted}}$, is an exponential weighted average determined for each link according to (7):

$$U_{\text{Predicted}} = \frac{C \cdot U_{\text{Short}} + U_{\text{Predicted}}}{C + 1},$$  (7)

where $C$ is the weight given to the short-term utilization over the long-term utilization. After $T$ cycles have elapsed, where $1/T$ is the maximum allowable switching rate, the router determines whether a given link's predicted utilization meets a specific threshold. By allowing thresholds at several different levels of $U_{\text{Predicted}}$, a finer-grain balance between energy savings, due to lowering the voltage and frequency, and latency penalty, due to mispredictions and voltage/frequency transitions, can be obtained.

Voltage regulators are required to step up or step down voltage in order to dynamically adjust voltage, and hence frequency. By following [11], using on-chip voltage regulators can reduce the transition time between voltage levels over standard off-chip voltage regulators.

Similar to [34], a DVFS algorithm was developed using (6) and (7). After $T$ cycles, the algorithm determines if DVFS should be performed on the link based on the predicted bandwidth requirements of future traffic. Depending on which threshold was crossed, if any, the router then determines whether or not to tune the voltage and frequency of the link. In order to prevent a direct multi-threshold jump, which would cause high delay, the voltage and frequency can step up once, step down once, or remain unchanged during one voltage/frequency transition. After each adjustment of the voltage/frequency pair on a given link, energy savings and latency penalty was determined. The energy of the link, $E_{\text{link}}$, was determined by (8):

$$E_{\text{link}} = \sum_T N_{\text{flits}} \cdot E_{\text{flit}} \cdot V_T^2,$$  (8)

where $N_{\text{flits}}$ is the number of flits over the period $T$, $E_{\text{flit}}$ is the energy per flit of the link, and $V_T$ is the DVFS-scaled voltage for the given period.

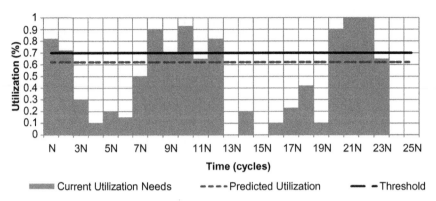

**Fig. 3.** DVFS link latency penalty determination.

The total energy of the link is summed over all switching periods within the entire simulation.

Latency penalty due to DVFS is composed of two main factors:

1. A misprediction penalty is caused when the adjusted voltage/frequency pair did not meet the bandwidth requirements of the traffic over the given switching interval. The bandwidth requirement of the link was obtained by viewing the current link utilization over a smaller window whose size was determined as the average latency of a flit in the non-DVFS network. An example of the misprediction penalty during a snapshot of the FFT benchmark can be seen in Fig. 3, where, each bar represents the link utilization over an $N$-cycle window. If the bar is higher than the threshold line, the bandwidth requirements of the flits traversed in that window were not met. This results in a latency penalty for the flits in that $N$-cycle window. This penalty can be considered as worst case, as it assumes that every flit is time critical, and a processor may be able to progress after the flit arrives.

2. Adjusting the voltage/frequency pair on the link causes a switching penalty. A transition delay of 100 ns was accounted for according to [11]. During this transition, the network conservatively does not attempt to send any information on the link.

To determine the appropriate switching window, the misprediction penalty and switch penalty of a given link in the presence of FFT traffic while varying $T$ is plotted in Fig. 4. A small switching window may catch data bursts, which do not represent a long-term trend of the benchmark's traffic. Consequently, widely varying short-term traffic utilizations, which

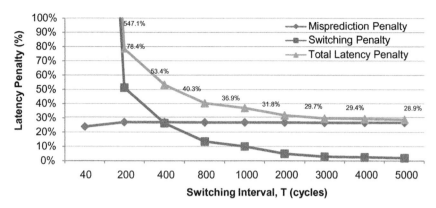

**Fig. 4.** Latency penalty versus switching window size of a link (FFT).

**Table 1** Voltage/frequency/threshold combinations.

| Voltage (V) | Frequency (GHz) | Threshold |
| --- | --- | --- |
| 1 | 2.5 | $\geq 0.9$ |
| 0.9 | 2.25 | $\geq 0.8$ |
| 0.8 | 2.0 | $\geq 0.7$ |
| 0.7 | 1.75 | $\geq 0.6$ |
| 0.6 | 1.5 | $\geq 0.5$ |
| 0.5 | 1.25 | $< 0.5$ |

can be seen in Fig. 3, will cause the voltage/frequency to change often. As seen in Fig. 4, for small switching windows, there is a large penalty due to frequently changing the voltage/frequency pair of the link. As the switching window widens, the switch penalty reduces drastically, while the overhead due to latency penalty increases slowly. A switching window size of $T = 5000$ was selected as the benefits of a larger window size beyond that were minimal.

From [34, 11], appropriate voltage/frequency pairs were obtained. Based on these pairs, required bandwidth thresholds were determined to be proportional to the scaled voltage/frequency pairs. The values can be seen in Table 1.

The voltage/frequency is scaled after $T$ cycles have elapsed in order to prevent large latency penalties due to frequency switching. This method will attempt to predict the next $T$ cycles correctly by tracking the utilization of the link over time. Figure 5 demonstrates how the link frequency tracks the utilization over an $H$-cycle window.

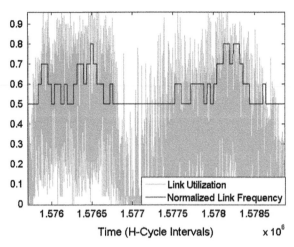

**Fig. 5.** Sample FFT frequency tracking.

## 4. PERFORMANCE EVALUATION

In this section the performance of the proposed WiMesh architecture is characterized by incorporating DVFS through detailed full system simulations. The GEM5 [35] platform is used, with Ruby and Garnet for carrying out detailed performance evaluations. GEM5 is a cycle accurate full system simulator. Ruby and Garnet allow the modification of the interconnection network within the GEM5 platform. Using these tools allows for the extraction of latency information of the network as well as detailed flit level information for realistic workloads. The NoC routers are synthesized from a RTL level design using 65 nm standard cell libraries from CMP ( http://cmp. imag.fr), using Synopsys™ DesignVision. The NoC routers are driven with a clock frequency of 2.5 GHz. Three SPLASH-2 [36] benchmarks, FFT, RADIX, LU, and the PARSEC benchmark, CANNEAL [37] are considered to study the latency, energy dissipation, and thermal characteristics of WiMesh.

Within the GEM5 platform, the benchmarks are run from beginning to the end, and statistics are obtained at the completion of each benchmark. A system of 64 alpha cores running Linux is considered. The cores are arranged on a 20 mm × 20 mm die. The width of all wireline links is considered to be the same as the flit size, which is 32.

## 4.1 Performance Metrics

As mentioned above, three parameters: latency, energy dissipation, and thermal profile are considered. Latency values were obtained through Garnet output. The average network latency per flit is considered as the relevant parameter. The latency penalty caused by DVFS was calculated as described in Section 3.3. The energy dissipations of network routers were obtained from the synthesized netlist by running Synopsys™ Prime Power. The energy dissipation of the wireline links were obtained through HSPICE simulations taking into account the specific lengths of each link based on the established connections in the 20 mm × 2mm die. Each wireless link can sustain a data rate of 10 Gbps [2]. The energy dissipation of each wireless link was found to be 0.33 pJ/bit, which is significantly less than even most efficient metal wires [2]. Energy savings by enabling DVFS on the wireline links was calculated as described in Section 3.3.

As temperature is closely related to the energy dissipation of the IC, the thermal profile depends on the energy dissipation of the NoC, which is quantified in Section 4.3. To quantify the effects of the WiMesh architecture on heat, its thermal profile is evaluated. The temperature profile of the WiMesh and D–WiMesh is obtained by using the HotSpot tool [38]. To further compare the characteristics of the thermal profiles of a particular region in the NoC, another important parameter that needs to be determined is the average communication density given by (9):

$$\rho_{\text{comm}} = \frac{1}{C} \left[ \sum_{i=1}^{N_{\text{router}}} N_{\text{fr}_i}/N_{\text{router}} + \sum_{j=1}^{N_{\text{link}}} N_{\text{fl}_j}/N_{\text{link}} \right], \qquad (9)$$

where $\rho_{\text{comm}}$ is the average communication density over the region of interest. $N_{\text{router}}$ is the number of routers in the region of interest, $N_{\text{link}}$ is the number of links in the region of interest, $N_{\text{fr}}$ and $N_{\text{fl}}$ are the number of flits traveled on each router and link in that region respectively, and $C$ is the total number of cycles executed. As communication density increases, temperature increases.

## 4.2 Latency Characteristics

In this subsection, the performance of the proposed WiMesh architecture is evaluated in terms of latency. Figure 6 shows the latency of the WiMesh and DVFS-enabled WiMesh versus the traditional mesh. As seen in Fig. 6, the reduction in latency of the WiMesh in presence of FFT, RADIX, LU, and CANNEAL traffics are 22.57%, 24.88%, 23.08%, and 23.75%, respectively

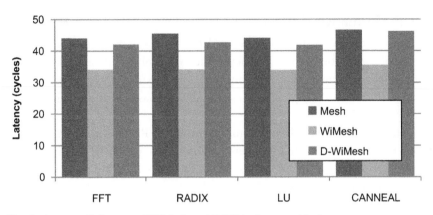

**Fig. 6.** Average flit latency of WiMesh and D-WiMesh versus Mesh.

compared to a conventional mesh. The latency savings of the WiMesh over the conventional wireline mesh was similar for all of the benchmarks considered. The latency savings is due to the wireless link placement that takes into account interactions among the cores depending on the traffic pattern in addition to the physical distance between them. By taking a wireless long-range link, multiple hops have effectively been bypassed that would have been taken with a traditional mesh-based design. By enabling DVFS on the wireline links of the WiMesh architecture, the latency penalty from dynamic adjustments to the voltage/frequency as well as mispredictions can balance with the latency improvements of the WiMesh architecture. The net latency of the D-WiMesh still is an improvement over the traditional wireline mesh. The latency savings as seen in Fig. 6 are 4.46%, 6.07%, 5.18%, and, 0.81% in presence of FFT, RADIX, LU, and CANNEAL traffics, respectively. This proves the benefit of the WiMesh architecture. The latency improvement introduced by the insertion of long-range wireless links is such that even in presence of DVFS, there is no overall latency penalty with respect to a conventional mesh NoC.

## 4.3 Energy Dissipation

In this subsection the energy dissipation characteristics of the WiMesh both with and without DVFS are evaluated, and compared with the conventional mesh. The total energy of all routers and links in the NoC was determined. The energy savings can be seen in Fig. 7. The total energy/cycle for the mesh running the FFT benchmark is 370 pJ/cycle, where the WiMesh dissipates

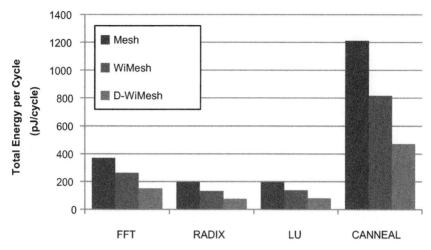

**Fig. 7.** Energy/cycle of WiMesh and D-WiMesh versus Mesh.

262 pJ/cycle. For RADIX traffic, the mesh dissipates 198 pJ/cycle while the WiMesh dissipates 131 pJ/cycle. The LU traffic running on mesh and WiMesh dissipate 196 pJ/cycle and 138 pJ/cycle, respectively. Finally, with CANNEAL traffic, the energy per cycle of the mesh and WiMesh were 1211 pJ/cycle and 817 pJ/cycle, respectively.

From Fig. 7 it can be seen that by appropriately placing the long-range wireless links according to the traffic pattern, energy savings is obtained. The wireless links handled 33%, 33%, 35%, and 33% of all traffic for the FFT, RADIX, LU, and CANNEAL benchmarks, respectively. This is a significant amount of traffic considering there are only 24 long-range wireless links whereas there are 112 wireline links. As a high amount of traffic is carried by the low power wireless links, the savings in energy dissipation is significant. Conversely, this still suggests that the majority of the traffic travels on the wireline links. To further reduce the energy dissipation of the flits traversing on wireline links, DVFS is enabled on them. From Fig. 7, it can be seen that by dynamically reducing the voltage and frequency of the wireline links in the WiMesh that the energy dissipation has reduced further in comparison to the traditional mesh. The D-WiMesh dissipates 151 pJ/cycle, 75 pJ/cycle, 81 pJ/cycle, and 471 pJ/cycle in presence of FFT, RADIX, LU, and CANNEAL traffics, respectively. This corresponds to 59.22%, 62.01%, 58.58%, and 61.11% energy savings for the respective benchmarks over the traditional mesh.

## 4.4 Thermal Profile

In this subsection the thermal profile of the WiMesh incorporating DVFS is evaluated. In Fig. 8, the hotspot regions forming in the mesh based on traffics can be seen. This phenomenon can be quantified further by obtaining the communication densities of the WiMesh and traditional mesh architectures in presence of the different traffics. The average communication density, shown in Fig. 9 was determined over the hotspot area of the network for each benchmark to analyze the thermal profile. As can be seen, there is a large reduction in communication density within the hottest area of the chip of the WiMesh. This directly relates to the thermal profile of the chip, as reduction in communication density will produce less heat.

To be more exact, the average link communication densities in WiMesh within the hotspot region of interest were reduced by 83.3%, 79.4%, 75.4%, and 88.65% for the FFT, RADIX, LU, and CANNEAL traffics, respectively, compared to the traditional mesh topology. The long-range links overwhelmingly take a lot of pressure away from the hotspot region by reducing the communication density by a significant amount. Similarly, savings

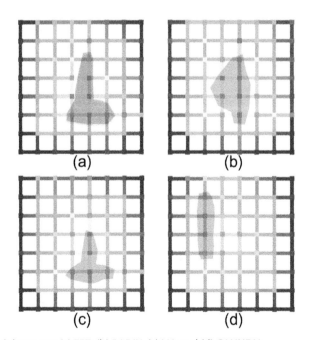

**Fig. 8.** Mesh hotspots: (a) FFT, (b) RADIX, (c) LU, and (d) CANNEAL.

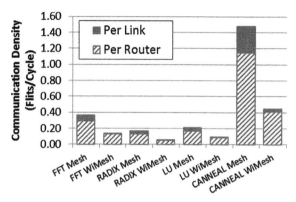

**Fig. 9.** Average communication density of hotspot region.

of average router communication densities were 56.4%, 61.3%, 49.7%, and 64.5% for the FFT, RADIX, LU, and CANNEAL traffics, respectively.

As explained in Section 4.1, the HotSpot tool is used to determine the thermal profile. Figure 10 depicts the temperature profiles of the conventional wireline mesh, WiMesh, and D-WiMesh for the different traffics tested.

The CANNEAL benchmark, with significantly higher communication density, provides the most noteworthy thermal results. Within the traditional mesh hotspot, the hottest router was determined to be 74.96 °C. This same router within the WiMesh architecture was reduced by 17.24–57.72 °C. This corresponds to a 23% reduction in temperature. This large temperature difference directly relates to the large reduction in communication density as mentioned earlier within the hotspot region. With the DVFS-enabled WiMesh, the energy of the wireline links has been further reduced, and the corresponding ports of the routers. By implementing DVFS on top of the WiMesh, the same hotspot router was reduced an additional 5.32 °C over the WiMesh. The hottest spot of the original mesh network has been reduced by 22.56 °C, corresponding to a 30.1% temperature decrease by implementing DVFS-enabled WiMesh. Similarly, the hottest spot of the mesh was reduced by 7.31 °C (13.5%), 3.23 °C (6.6%), and 3.42 °C (6.9%) for the FFT, RADIX, and LU benchmarks, respectively. The massive savings from the CANNEAL benchmark can be explained as the hottest part was 74.96 °C, over the other benchmarks having hottest spots of 53.99 °C, 49.2 °C, and 49.92 °C. With the chip heat over 20 °C hotter when running the CANNEAL benchmark, the true benefits of the WiMesh architecture can be seen when the chip incurs significant traffic in a localized region

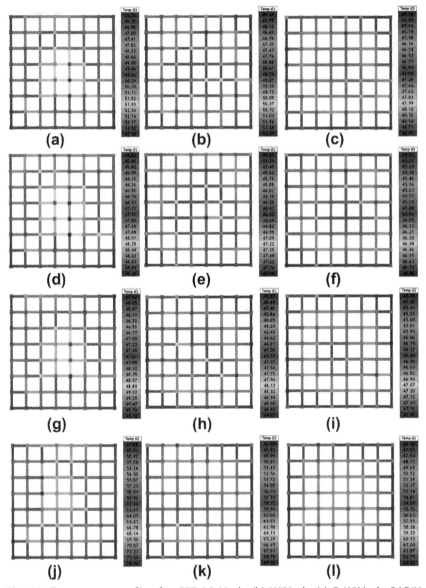

**Fig. 10.** Temperature profiles for: FFT (a) Mesh; (b) WiMesh; (c) D-WiMesh, RADIX; (d) Mesh; (e) WiMesh; (f) D-WiMesh, LU; (g) Mesh; (h) WiMesh; (i) D-WiMesh, CANNEAL; (j) Mesh; (k) WiMesh; and (l) D-WiMesh.

of the chip. Implementing DVFS further decreases power consumption chip-wide, as can be seen between the WiMesh and D–WiMesh configurations of Fig. 10.

# 5. PROCESSOR-LEVEL DVFS

In this section, the DVFS mechanism applied to the processors is presented. It complements the network-level DVFS investigation elaborated above. A hybrid algorithm is proposed, which makes use of compiling information to achieve runtime frequency and voltage scaling. In this complier-oriented DVFS algorithm, the program behavior is verified rather than predicted. In order to evaluate the algorithm, GEM5 is used to simulate a 64-core system with homogenous Alpha processing cores. Cores are assumed to be independent of each other so that the state (i.e. frequency and voltage) of each processor can be adjusted separately. GEM5 is an event driven computer system simulator platform, which provides us with statistics of the target program running on the above system, including timing and hardware behavior. The corresponding runtime power is generated by feeding these statistics into McPAT [39].

## 5.1 DVFS Algorithm

The proposed algorithm inserts directory flags in the multi-core program running with GEM5. These flags indicate the beginnings of program sections, which could possibly cause CPU idle periods. When the multi-core program is executed, the DVFS algorithm will not be triggered unless a flag is detected. By combining offline information and runtime behavior, the aim is to adjust frequency and voltage precisely.

No matter how the program halts (i.e. lock acquisition, lock release, barrier, memory access, etc.), at the instruction level, the program would almost always halt on a *load* instruction. Since the goal is to build a fine-grained DVFS scheduling algorithm, focus is given to dealing with *load* instruction, which can potentially cause CPU idle cycles. The approach is divided into two phases:

*Phase I (static approach):* Insert directory flags in front of proper *load* instructions during compile time.

*Phase II (dynamic approach):* The DVFS mechanism is activated only if the processors detect the flags during execution of the program.

In phase I, only those *load* instructions that have data reference issues are of concern. Due to the pipelined processor architecture, if the required data in *load* instruction is used far away from the *load*, it will not cause any CPU idle time even if the data is not a cache resident. Based on the above experimental setup, the following estimations are made: L1 cache hit takes 8 CPU cycles to fetch the data, L2 cache hit takes 32 CPU cycles to fetch the data, and it takes 200 cycles to fetch data from main memory if the data is ready there. Due to a high cache hit rate, most of the data will be fetched from L1 cache, which takes 8 cycles. If the required data will be used more than 8 cycles later than the corresponding *load* instruction, the possibility that the program would be halted by this data is significantly small. Therefore, only a flag is inserted in front of such *load* that the required data will be used within 8 cycles.

On the other hand, in phase II, when a flag is detected, the DVFS procedure will keep track whether the data is fetched. If the data is not back within 32 cycles, a L2 cache miss is detected. Similarly, a main memory miss, which could be caused by lock or barrier in the program, will be detected if data is not back within 200 cycles. Once a miss is detected, frequency and voltage will be decreased by one stage. This state transition takes 100 ns to complete [11]. If more than one miss are detected during the transition, the latter ones would not cause a frequency decrease but the number of misses will be counted in *NumberOfFLAG*. On contrary, once the required data is fetched and no more data miss is detected, frequency and voltage will be increased by one stage; otherwise the CPU state remains the same even if data is back. It is reasonable because although the previous data has come back, new potential idle time is detected. In order to waste as less time as possible, processing is done during state transition time. Algorithm 1 describes the steps involved in phase II of the proposed DVFS mechanism. In each decoding stage of an instruction the flags are detected and a timer is set up for data fetching. Since both a cache miss and a main memory miss can cause a frequency decrease, the variable *loadstep* is used to indicate the status of load.

**Algorithm 1.**   Compiler-Oriented DVFS Algorithm Phase II

```
NumberOfFLAG = 0;// number of pending code sections
loadstep = 0;// indicates load status: initially no load
for each DECODE stage do
   if (OP == FLAG) then
      increase NumberOfFLAG by 1
      if (timer == 0 and loadstep == 0) then
         set timer to 32
         loadstep = 1// try to load from cache
      end if
   end if
   do regular decoding work
end for
```

```
for each cpu cycle do
    if required data is back then
        decrease NumberOfFLAG by 1
        loadstep = 0// load finished
        reset timer to 0
        if (NumberOfFLAG == 0) then
            increase power state
        end if
    else
        if (timer == 0) then
            decrease power state
            if (loadstep == 1) then
                set timer to 200
                loadstep = 2// try to load from main memory
        end if
    else
        decrease timer
        end if
    end if
end for
```

## 5.2 Performance Evaluation

The performance of the proposed DVFS algorithm using the above system configurations is evaluated in presence of FFT, LU, RADIX, and CAN-NEAL as tested earlier by the network. The coarse-grained execution time breakdown is shown in Fig. 11.

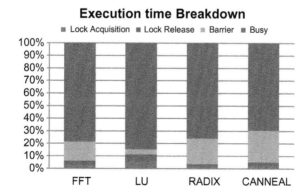

**Fig. 11.** Coarse-grained execution time breakdown of benchmarks.

**Table 2** CPU level DVFS simulation results.

| | FFT | | LU | | RADIX | | CANNEAL | |
|---|---|---|---|---|---|---|---|---|
| | Execution Time (s) | Energy (J) | Execution Time (s) | Energy (J) | Execution Time (s) | Energy (J) | Execution Time (s) | Energy (J) |
| Without DVFS | 0.003172 | 0.0881 | 0.092254 | 2.96353 | 0.064782 | 1.72233 | 0.143129 | 3.923509 |
| With DVFS | 0.003987 | 0.10678 | 0.103724 | 2.85984 | 0.073981 | 1.55443 | 0.150872 | 3.195228 |
| Penalty/savings (%) | 25.69 | 21.20 | 12.43 | −4.50 | 14.20 | −9.75 | 5.41 | −18.56 |

The main simulation results are shown in Table 2. The last row of Table 2 represents the penalty or savings in both execution time and energy, where positive numbers mean that a penalty is paid while negative numbers mean that there is some savings. There are two main factors that affect the performance of the proposed DVFS algorithm: computing intensity and overall execution time. As shown in Fig. 11, CANNEAL is much less compute-intensive than other three benchmarks, and thus it is able to save nearly 20% of energy with only 5.41% execution time penalty. Correspondingly, in RADIX and LU the approach achieves 9.75% and 4.50% savings of energy with latency penalty 14.20% and 12.43%, respectively. Meanwhile, when the overall execution time shrinks to a certain extent, it influences the algorithm's performance. FFT is a good example to illustrate this. From Fig. 11 it can be concluded that the algorithm will be activated a considerable number of times in FFT similarly like RADIX, and the amount of energy savings of FFT and RADIX should be similar on computing intensity. However, since the execution time of FFT is just 0.003172 s, the 100 ns transition time plays an important role. Frequent state transition makes the processors work in transition mode almost all the time when frequency and voltage do not match, i.e. high voltage with low frequency, which cause plenty of energy waste and execution slowdown. It results in paying more than 20% penalties in both execution time and energy.

## 6. COMPLETE THERMAL PROFILE

In this section the overall thermal profile of the whole chip considering DVFS in both the network and processor levels is presented. The importance of co-existent network and processor level DVFS is to fight the soaring temperatures caused by the high energy density in the multi-core chip. By implementing this dual level DVFS the thermal profile of the chip can be significantly improved. As seen in Fig. 12, the temperature of the whole system is extremely high. The chip will not be able to sustain this high temperature and the reliable operation will be severely compromised.

One simple method to keep the temperature within sustainable limits would be to assign the processes that are dissipating the most energy to the processors along the edge of the chip. By doing this, in Fig. 13 temperature reductions of 3.1%, 27.1%, 22.8%, and 0.5% can be seen for FFT, LU, RADIX, and CANNEAL benchmarks, respectively. There are three separate scenarios happening here. For FFT, the processes are already performing similar amounts of work as the temperatures among the cores are spread across

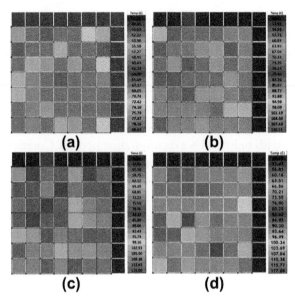

**Fig. 12.** Mesh-based thermal profiles (a) FFT, (b) LU, (c) RADIX, and (d) CANNEAL.

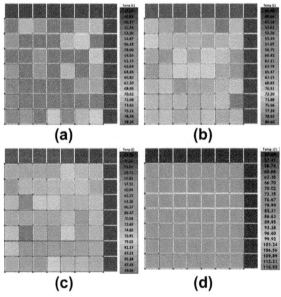

**Fig. 13.** Mesh based, strategic process placement, thermal profiles (a) FFT, (b) LU, (c) RADIX, and (d) CANNEAL.

the chip somewhat evenly. Due to this, strategically placing the processes with the highest energy dissipation along the chip edge will not help the thermal profile immensely. The second type, which can be seen in the LU and RADIX benchmarks, helps the thermal profile drastically. Due to several processes dominating the energy dissipation of the chip, moving them near the chip edge helps to reduce the temperature of the chip by as much as 27%. The worst case, seen in CANNEAL, is that where the significance in the energy dissipation is network based. Moving processes will not give a large benefit when the network is the main contributor of chip temperature. This is one reason that alleviating the network through WiMesh and DVFS are important.

Furthermore, by applying the dual level DVFS to the whole system, the temperatures can be reduced to much lower, sustainable levels for all the benchmarks considered. Figure 14 displays the improvement in the thermal profile from the strategically placed processes on the mesh to the dual-DVFS enabled strategically placed processes on the WiMesh. Here the high energy dissipation within the network running CANNEAL traffic has been resolved. Another noticeable outcome is that the hottest spot within the chip has been further reduced by 0.6%, 11.7%, 11.2%, and 27.1% for FFT, LU,

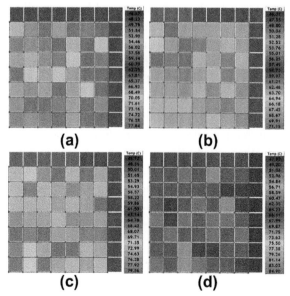

**Fig. 14.** WiMesh-based dual-DVFS, strategic process placement, thermal profiles (a) FFT, (b) LU, (c) RADIX, and (d) CANNEAL.

RADIX, and CANNEAL benchmarks respectively. With modest savings in energy for LU and RADIX, and even an increase in overall energy in FFT when running the processor level DVFS, the processes seem to balance out their energy dissipation better, improving overall thermal profile. In presence of CANNEAL traffic, the dual-DVFS algorithms significantly help the thermal profile.

The latency improvement from the network part of the D-WiMesh is minimal, but the performance of the whole system will be bottlenecked by the execution time penalty introduced by the processing cores. The overall latency increase for the various benchmarks can be set to worst case as the values from Table 2. From this, the benchmarks had an overall latency penalty of 25.7%, 12.4%, 14.2%, and 5.4% for the FFT, LU, RADIX, and CANNEAL respectively. Thus, in some instances, such as FFT, processor level DVFS is not worth the thermal benefits due to the 25.7% latency increase. However, a tradeoff of a 12.4% and 14.2% latency increase may be reasonable for a 11.7% and 11.2% thermal decrease. One clear advantage is with CANNEAL. Here, a temperature hotspot reduction of 27.1% results only in a latency increase of 5.4%.

## 7. CONCLUSION

In this chapter, it has been demonstrated how a small-world DVFS-enabled wireless NoC improves both the energy and thermal profile of a multi-core chip. By adopting a small-world interconnection infrastructure, where long-distance communications will be predominantly achieved through high-performance specialized single-hop wireless links, communications can be made significantly more energy efficient. The low power and long-range wireless links carry a significant percentage of the overall traffic, and hence the temperature hotspot regions in the system are able to be reduced. To further extend the energy savings, implementing link level DVFS on wireline links, enable significant reductions to energy, and hence reduce the overall chip temperature more. By adding processor level DVFS and assigning processes appropriately among the cores, additional savings to energy and heat are obtained.

## LIST OF ABBREVIATIONS

CNT          Carbon Nanotube
CVD          Chemical Vapor Deposition

| DVFS | Dynamic Voltage and Frequency Scaling |
|------|----------------------------------------|
| D-WiMesh | DVFS-enabled WiMesh |
| EM | Electromagnetic |
| FDM | Frequency Division Multiplexing |
| FFT | Fast Fourier Transform |
| G-lines | Global lines |
| MEMS | Micro-electro-mechanical Systems |
| MWCNT | Multi-Walled Carbon Nanotube |
| NoC | Network-on-Chip |
| RTL | Register Transfer Logic |
| SA | Simulated Annealing |
| UWB | Ultra Wideband |
| VFI | Voltage Frequency Island |
| WiMesh | Wired Mesh with Wireless Shortcuts |
| WiNoC | Wireless NoC |

# REFERENCES

[1] U.Y. Ogras, R. Marculescu, It's a small world after all: NoC performance optimization via long-range link insertion, IEEE Transactions on Very Large Scale Integration (VLSI) Systems 14 (7), 2006, 693–706.

[2] A. Ganguly et al., Scalable hybrid wireless network-on-chip architectures for multi-core systems, IEEE Transactions on Computers 60 (10), 2011, 1485–1502.

[3] S. Deb et al., Enhancing performance of network-on-chip architectures with millimeter-wave wireless interconnects, in: Proceedings of IEEE International Conference on ASAP, Rennes, France, 2010, pp. 73–80.

[4] R. Marculescu et al., Outstanding research problems in NoC design: system, microarchitecture, and circuit perspectives, IEEE Transactions on Computer-Aided Design of Integrated Circuits and Systems 28 (1) (2009) 3–21.

[5] A. Kumar, L.-S. Peh, P. Kundu, N.K. Jha, Toward ideal on-chip communication using express virtual channels, IEEE Micro 28 (1), 2008, 80–90.

[6] D. Zhao, Y. Wang, SD-MAC: design and synthesis of a hardware-efficient collision-free QoS-Aware MAC protocol for wireless network-on-chip, IEEE Transactions on Computers 57 (9), 2008, 1230–1245.

[7] S.B. Lee et al., A scalable micro wireless interconnect structure for CMPs, in: Proceedings of ACM Annual International Conference on Mobile Computing and Networking, Beijing, CN, ACM, 2009, pp. 217–228.

[8] D. DiTomaso et al., iWise: Inter-router wireless scalable express channels for network-on-chips (NoCs) architecture, in: Proceedings of Annual Symposium of High Performance Interconnects (HOTI), 2011, pp. 11–18.

[9] S. Garg, D. Marculescu, R. Marculescu, Technology-driven limits on run-time power management algorithms for multi-processor systems on chip, ACM Journal on Emerging Technologies in Computing Systems, 8 (4), 2012.

[10] J. Howard et al., A 48-Core IA-32 Processor in 45 nm CMOS using on-die message-passing and DVFS for performance and power scaling, IEEE Journal of Solid-State Circuits 46 (1) (2011).

[11] W. Kim, M. Gupta, G.-Y. Wei, D. Brooks, System level analysis of fast, per-core DVFS using on-chip switching regulators, in: Proceedings of the International Symposium on High Performance Computer Architecture, 2008, pp. 123–134.

[12] J. Luo, N.K. Jha, Li-Shiuan Peh, Simultaneous dynamic voltage scaling of processors and communication links in real-time distributed embedded systems, IEEE Transactions on Very Large Scale Integration (VLSI) Systems 15 (4) (2007).

[13] W. Jang, D. Ding, D. Pan, A voltage-frequency island aware energy optimization framework for networks-on-chip, in: Computer-Aided Design, ICCAD 2008.

[14] U. Ogras, R. Marculescu, D. Marculescu, Variation-adaptive feedback control for networks-on-chip with multiple clock domains, in: Proceedings of the 45th Annual Conference on Design Automation 2008, pp. 614–619.

[15] K. Niyogi, D. Marculescu, Speed and voltage selection for GALS systems based on voltage/frequency islands, in: Proceedings of the 2005 Conference on Asia South Pacific Design Automation, 2005, pp. 292–297.

[16] S. Dighe et al., Within-die variation-aware dynamic-voltage-frequency scaling core mapping and thread hopping for an 80-core processor, in: Solid-State Circuits Conference Digest of Technical Papers (ISSCC), 2010 IEEE International, IEEE, 2010, pp. 174–175.

[17] E. Beigne et al., An asynchronous power aware and adaptive NoC based circuit, IEEE Journal of Solid-State Circuits 44 (4), 2009, 1167–1177.

[18] J. Donald, M. Martonosi, Techniques for multicore thermal management: classification and new exploration, in: Proceedings of the 33rd International Symposium on Computer Architecture (ISCA'06), 2006, pp. 78–88.

[19] L. Shang et al., Temperature-aware on-chip networks, in: IEEE Micro: Micro's Top Picks from Computer Architecture Conferences, 2006, pp. 130–139.

[20] S. Deb, A. Ganguly, P. Pande, B. Belzer, D. Heo, Wireless NoC as interconnection backbone for multicore chips: promises and challenges, IEEE Journal on Emerging and Selected Topics in Circuits and Systems 2 (2) (2012).

[21] A. Ganguly et al., Scalable hybrid wireless network-on-chip architectures for multi-core systems, IEEE Transactions on Computers (TC) 60 (10) (2011) 1485–1502.

[22] B.A. Floyd et al., Intra-chip wireless interconnect for clock distribution implemented with integrated antennas, receivers, and transmitters, IEEE Journal of Solid-State Circuits 37 (5) (2002) 543–552.

[23] G.W. Hanson, On the applicability of the surface impedance integral equation for optical and near infrared copper dipole antennas, IEEE Transactions on Antennas and Propagation 54 (12) (2006) 3677–3685.

[24] K. Kempa et al., Carbon nanotubes as optical antennae, Advanced Materials 19 (3), 2007, 421–426.

[25] P.J. Burke et al., Quantitative theory of nanowire and nanotube antenna performance, IEEE Transactions on Nanotechnology 5 (4) (2006) 314–334.

[26] Y. Huang et al., Performance prediction of carbon nanotube bundle dipole antennas, IEEE Transactions on Nanotechnology 7 (3) (2008) 331–337.

[27] Y. Zhou et al., Design and fabrication of microheaters for localized carbon nanotube growth, in: Proceedings of IEEE Conference on Nanotechnology, 2008, pp. 452–455.

[28] A. Shacham et al., Photonic network-on-chip for future generations of chip multiprocessors, IEEE Transactions on Computers 57 (9) (2008) 1246–1260.

[29] M. Freitag et al., Hot carrier electroluminescence from a single carbon nanotube, Nano Letters 4 (6) (2004) 1063–1066.

[30] T.S. Marinis et al., Wafer level vacuum packaging of MEMS sensors, in: Proceedings of Electronic Components and Technology Conference, 2005, pp. 1081–1088.

[31] P. Bogdan, R. Marculescu, Quantum-like effects in network-on-chip buffers behavior, in: Proceedings of IEEE Design Automation Conference, DAC, 4–8 June, 2007, pp. 266–267.

[32] S. Kirkpatrick et al., Optimization by simulated annealing, Science, New Series 220 (4598) (1983) 671–680.

[33] B.G. Lee et al., Ultrahigh-bandwidth silicon photonic nanowire waveguides for on-chip networks, IEEE Photonics Technology Letters 20 (6), 2008, 398–400.
[34] L. Shang, L.-S. Peh, N. Jha, Power-efficient interconnection networks: dynamic voltage scaling with links, Computer Architecture Letters 1 (2002) 6.
[35] N. Binkert et al., The GEM5 simulator, ACM SIGARCH Computer Architecture News 39 (2), 2011, 1–7.
[36] S.C. Woo et al., The SPLASH-2 programs: characterization and methodological considerations, in: Proceedings of Annual International Symposium on Computer Architecture, Santa Margherita Ligure, IT, 1995, pp. 24–36.
[37] C. Bienia, Benchmarking Modern Multiprocessors, Ph.D. Thesis, Princeton University, January 2011.
[38] K. Skadron et al., Temperature-aware microarchitecture, in: Proceedings of the International Symposium on Computer Architecture, 2003, pp. 2–13.
[39] S. Li, J.H. Ahn, R.D. Strong, J.B. Brockman, D.M. Tullsen, N.P. Jouppi, McPAT: an integrated power, area, and timing modeling framework for multicore and manycore architectures, in: Proceedings of MICRO'09, 12–16 December, 2009, New York, NY, USA.

## ABOUT THE AUTHORS

**Jacob A. Murray** is working towards his PhD in Electrical and Computer Engineering at the Department of Electrical Engineering and Computer Science, Washington State University. His PhD research interests include sustainable and low-power on-chip interconnection networks. Before this, he received his bachelor's degree in Computer Engineering at Washington State University in 2010. He has been a Harold Frank Entrepreneur and participated as one of five undergraduate finalist teams in the 2010 National Collegiate Inventors Competition. He is a member of Tau Beta Pi, the national engineering honors society, and a student member of the IEEE.

**Teng Lu** received his bachelor's degree in Computer Science and Technology from Beijing University of Posts and Telecommunications, China and his master's degree in Telecommunications and Computer Engineering from London South Bank University, U.K. He is currently pursuing his doctoral studies in the area of Computer Science at Washington State University. His research areas include compiler directed dynamic voltage and frequency scaling and power aware scheduling.

**Partha Pratim Pande** received the M.S. degree in computer science from the National University of Singapore and the Ph.D. degree in electrical and computer engineering from the University of British Columbia, Vancouver, BC, Canada. He is an Associate Professor at the School of Electrical Engineering and Computer Science, Washington State University, Pullman. His current research interests are novel interconnect architectures for multicore chips, on-chip wireless communication networks, and hardware accelerators for biocomputing. He has more than 60 publications on this topic in reputed journals and conferences. He is the Guest Editor of a special issue on sustainable and green computing systems for ACM Journal on Emerging Technologies in Computing Systems. Dr. Pande currently serves on the Editorial Board of IEEE Design and Test of Computers and Sustainable Computing: Informatics and Systems (SUSCOM). He also serves in the program committee of many reputed international conferences.

**Behrooz A. Shirazi** is the Huie-Rogers Chair Professor and the Director of the School of Electrical Engineering and Computer Science at Washington State University. Dr. Shirazi has conducted research in the areas of sustainable computing, pervasive computing, software tools, distributed real-time systems, and parallel and distributed systems over the past eighteen years. He is currently serving as the Editor-in-Chief for Special Issues for the Pervasive and Mobile Computing (PMC) Journal and the Sustainable Computing (SUSCOM) Journal. He has served on the editorial boards of the IEEE Transactions on Computers and Journal of Parallel and Distributed Computing in the past. He is a co-founder of the International Green Computing Conference.

CHAPTER FOUR

# Smart Grid Considerations: Energy Efficiency vs. Security

**Andreas Berl, Michael Niedermeier, and Hermann de Meer**
University of Passau, Computer Networks and Computer Communications, Innstr. 43D-94032 Passau, Germany

## Contents

## Abstract

The Smart Grid is expected to increase the efficiency of the current power grid, to cope with volatile power production based on renewable resources, to reduce the need for fossil-based energy resources, and to guarantee the stability of power supply. To achieve these objectives, today's power grid is enhanced by information and communication technology to increase the information flow and to enable a sophisticated power production and power demand management. However, as the power grid is extended to a network of networks, it does not only become smarter, but also more vulnerable to security threats. This chapter discusses the current status and future developments of the Smart Grid and its challenges. Enhancements in terms of energy efficiency and new energy management approaches are covered as well as novel security challenges in different parts of the Smart Grid architecture. In short, this chapter analyzes some of the most striking risks and threats concerning the new Smart Grid infrastructure and discusses interdependencies between energy efficiency and security in the Smart Grid.

# 1. INTRODUCTION

The combined volatility of both, power supply and power demand, creates a growing problem that needs to be solved by the Smart Grid. On the one hand, volatile power demands lead to peak loads that need to be satisfied by inefficient peaking power plants, such as generators powered by fossil fuels. On the other hand, the increasing power production based on renewable sources tends to be subject to uncontrollable factors, e.g., wind or sunlight, and power needs to be consumed as available. The Smart Grid needs to address this imbalance of power supply and demand and maintain the power grid in a stable state.

To achieve these goals, the Smart Grid realizes a complex energy management that tries to reshape power production and demand to fit to the dynamic availability of regenerative energy sources. On the user side, new devices such as smart meters and smart appliances are used to achieve energy management. Advanced infrastructures, as enhanced supervisory control and data acquisition (SCADA) systems are enablers of energy management at the supplier side. In households, smart appliances will reduce or delay power demand while power is expensive (during times with peak load power consumption) and consume more power while it is cheap (when regenerative energy sources are available). As an example, e-cars are able to vary time and rate at which their rechargeable batteries are loaded in reaction to varying power availability and prices. Similarly, heating systems and air conditioners are able to dynamically adjust their demand.

To enable the envisioned energy management in the Smart Grid, information on current power consumption and the availability of power needs to be exchanged between power consumers and power suppliers. There-fore, new information flows need to be established and Smart Grid devices need to be interconnected. This interconnection of grid technology with information and communication technology (ICT), however, leads to novel security challenges in the formerly isolated power grid. One of the key chal-lenges and major obstacles in the widespread deployment of the Smart Grid is user privacy. The Smart Grid relies heavily on the usage of smart meter infrastructures for pricing and feedback purposes. Fine granular readings of power consumption are transmitted to power suppliers. These power con-sumption profiles are used to enable a precise prediction of power demand in order to control power production accordingly. Power consumption pro-files, however, allow for the creation of usage profiles of specific persons, households, or companies. Such profiles can be analyzed to identify the per-sonal behavior of users or to evaluate the business activity in enterprises. Therefore, measures have to be taken to ensure the required level of privacy in these areas. Additionally, the ICT-based management approach used in the Smart Grid brings forth new attack motives (discussed in more detail in Section 3.1.2), inducing new forms of security threats [1], such as:

- Energy theft by customers has always been a problem and is still causing massive financial losses to power suppliers. Incentives for energy theft may raise with the possibility of ICT-based manipulations.
- Cyber terrorists may try to shut down important parts of the grid to force directed or undirected blackouts.
- Organized crime may try to get access to critical control systems to threaten or blackmail power suppliers.

One of the main problems in this context is that the Smart Grid is a crit-ical infrastructure, where even small blackouts can cause significant social, economical, and ecological damage.

The remainder of this chapter is structured as follows: Section 2 describes the current status and future challenges of the power grid and a possible architecture of the future Smart Grid. Section 3 introduces security goals of the Smart Grid and identifies possible attackers. Furthermore, a deeper discussion on privacy and security is done in the context of smart house-holds and Smart Grid control systems. Section 4 discusses interdependencies between energy efficiency and security and evaluates energy efficiency as a conflicting goal with security. Section 5 discusses other work that is related to the topic of this chapter and Section 6 concludes this chapter.

## 2. SMART GRID

This section discusses the current status of the power grid in Section 2.1, power generation in Section 2.2, challenges that need to be solved by the Smart Grid in Section 2.3, and a possible Smart Grid architecture in Section 2.4.

### 2.1 Today's Power Grid

Many power grid systems used today are the same since the time of their creation. Just as an example, the American electric power system was created a century ago, while the European system has already some 50 years of age.[1] Operation methods have not changed until now, however, the number of consumers and the electricity demand has increased significantly over time. The worldwide electric power generation will double from about 17.3 trillion kW h in 2005 to 33.3 trillion kW h by 2030 [2]. In an electric power system [3], high voltage electricity is delivered from generating stations through an electric transmission and distribution system, and converted into manageable voltage levels to be used by customers. The North American power grid (often called "largest machine in the world") includes over 9000 generating stations and 700,000 miles of high voltage transmission lines, of which 200,000 miles operate at 230 kV or more. 1,000,000 miles of distribution lines are owned by over 3000 different utility entities and electricity is delivered to more than 334 million people [4]. Electricity is distributed through above- and underground wires, and the utility entities (or power companies) attempt to match up production to demand, in order to keep the system in balance. Transformers and mechanical breakers regulate the electricity flow. When the electricity levels are too high, overwhelming the lines is avoided by stopping the flow. After power generation is performed at power stations, the next step is to transform the power at transmission substations from medium voltage (15–50 kV) to high voltage (138–765 kV) with alternating current [5]. Close to the consumers, the power is stepped-down to lower voltages (10–34.5 kV), where it leaves the transmission system and enters the distribution system. The distribution system has the task of delivering power from the transmission system to consumers. Most homes receive power at 120 V (the standard in Europe is 240 V), where it is converted to a lower voltage to be used in appliances [6]. System operators continually monitor and control power generation and operation of the transmission and

---

[1] History of American electric power: http://www.aep.com/about/history/.

distribution grid in order not to overload the system and detect any anomaly or outage. This is an arduous task as in many cases the power supplier depends on customers, who should alert problems in the power supply. The power supplier tracks down the problem and sends a crew to fix it manually [7]. Moreover, North America's gigantic power system is increasingly outdated and overburdened, which leads to problems like the one registered at August 14th 2003, when power line failures caused a black out in the northern and eastern USA and Canada, knocked out power to approximately 5 million people, covered more than 9000 miles$^2$, caused three deaths and closed 12 major airports, finally causing a $6 billion loss in economic revenue, all in less than 48 h. But beyond power line failures, a determining factor was the struggle of system operators in their efforts to monitor the grid. Unfortunately, system operators did not have adequate tools in place to monitor, analyze, and control all relevant events, only limited real-time synchronized data was available. Therefore, operators were not able to detect the cascading effect fast enough. Furthermore, when actions could be taken to prevent the blackout spreading, the local utility's managers had to contact the regional system operator by phone in order to know what was happening on their own wires. While at the very same time, the failure spread to neighboring regions [8].

## 2.2 Power Mix

Although there were great efforts to achieve energy efficiency by reducing energy consumption in all kinds of areas, the worldwide energy consumption is increasing from year to year [9]. Traditionally, a threefold power mix was produced to satisfy power demand:

1. *Base load power generation:* The inflexible part of power demand is covered by power plants that are able to deliver high amounts of cheap power, as nuclear or coal power plants. The amount of generated power is relatively stable and the responsiveness to variations in power demand is very low.

2. *Medium load power generation:* The flexible part of power demand is covered by power plants that are able to deliver relatively high amounts of power, where the power production can be adapted to the varying demand. Power demand, may, e.g. be higher in the morning, highest at noon, and lower in the afternoon. Although these variations can be predicted to a certain limit, base load power generation is not able to cope with them. Therefore, power plants as combined gas and steam are used, which are more expensive than base load power generation, but also more responsive to power demand variations.

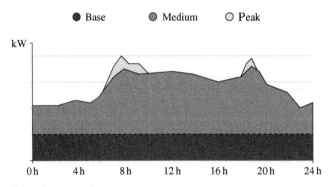

**Fig. 1.** Traditional power mix.

**3.** *Peak load power generation:* The most flexible and most unpredictable part of power demand is covered by pumped–storage hydroelectricity, gas turbines, or diesel generators, which are highly responsive to variations in power demand. However, fossil-based power generation is ecologically and economically expensive and also pumped-storage hydroelectricity has essential impact to the environment and is discussed controversially.

The more responsive power generation has to be, the more expensive it becomes, and the higher its impact on the environment is. The described traditional power mix is illustrated in Fig. 1 [10]. The X-axis shows a day in hours and the Y-axis shows a trend of power generation in kW. It can be observed that there is a constant base load power generation at the bottom of the graph, a more flexible medium load power generation, and there are two peaks (caused by unpredicted power demand) that need to be covered by peak load power generation.

Lately, an increasing amount of power based on renewable energy sources is fed into the power grid: On the one hand, there are world-wide efforts to reduce the emission of greenhouse gases (e.g. $CO_2$) to prevent further global warming. On the other hand, there are attempts to significantly reduce the usage of fossil-based power generation and nuclear power resources. Such resources are limited, fossil-based power resources are $CO_2$-intense, and nuclear power is controversial due to its safety (e.g. catastrophes of Chernobyl and Fokushima) and its massive long-term costs and dangers. Renewable energy sources cause no $CO_2$ emissions and do not waste limited resources. However, renewable energy sources are subject to uncontrollable factors such as wind or sunlight and need to be consumed as available. Figure 2 [10] depicts the volatility of power production. It can be seen that the prediction of a sufficient amount of medium load generation becomes highly difficult, due to the fluctuations in the availability of renewable energy sources.

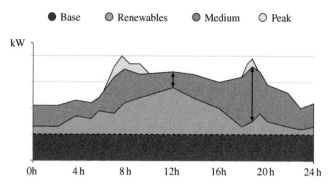

**Fig. 2.** Power mix including renewable energy sources.

## 2.3 Challenges

Power suppliers need to satisfy volatile power demand (foreseen and unforeseen power consumption) and to integrate the highly volatile production of power based on renewable sources. This volatility, however, has impacts on power grid frequencies, voltages, and component performance [11]. Despite of fluctuations in power demand and production, the power grid needs to provide a sufficient level of power quality and to maintain the power grid in a stable state.

The definition of power quality depends on the different perspectives of power generation, transmission, and consumption: Power quality from the generation perspective can be defined as "*the generator's ability to generate power at* 60 Hz *with little variation.*" From transmission and distribution perspective power quality can be defined as "*a nominal voltage staying within* ±5%." And finally, from the consumer perspective, power quality can be defined as "*voltage, current, and frequency that do not cause failure or misoperation of end-user's equipment*" [12]. Power quality can be seen as a delivery free of disruptions or disturbances. This delivery of high quality power is needed to keep commercial and industrial entities operational and working productively. However, there exist several factors that affect power quality, e.g. over/under voltages, voltage sags, outages, harmonic distortions, voltage swells, electrical noise, impulses or spikes, or flickers. The problem of insufficient power quality leads to fluctuations in power supply that used to cause losses of 15–24 billion dollars per year in the USA alone [13].

Another major issue of the future Smart Grid will be to keep the grid available [14]. This means, the power frequency (depending on power production and demand) needs to be kept within certain limits to keep the power grid stable. As an example, in Germany, the normal frequency is 50.2 Hz.

According to the Union for the Co-ordination of Transmission of Electricity (UCTE), who is responsible for the power grid in wide parts of Europe, the grid may become unstable if the frequency drops below 49.2 Hz or exceeds 50.8 Hz. Gaps between power supply and power demand heavily impact power quality and the stability of the grid. Due to the increasing proportion of power based on renewable sources, it becomes increasingly difficult and costly to minimize such gaps and to provide uninterrupted power to consumers [15]. There are two main challenges that have a major impact on power quality and grid stability:

1. *Shortage of power:* Shortage of power is caused by power demand that is not fully covered by power generation. On the one hand, there may be foreseen or unforeseen peaks in power demand, e.g., caused by special events (football match) or the weather (increased heating due to cold weather). On the other hand, there may be a sudden drop in the availabilty of renewable energy sources due to clouds or missing wind. During a shortage of power, economically and ecologically expensive peak load power generation needs to be activated or energy needs to be bought from neighboring countries. If this fails the grid becomes unstable.

2. *Surplus of power:* A surplus of power is caused if more power is produced than needed. This situation can be caused by a sudden increase in the availability of renewable energy sources or a limited demand of power (e.g. on sunny Sundays or holidays, when industry is not working and people are out in the sun). In this case it is possible to turn off parts of the power generation that is based on renewable sources (not all of this production is controllable) and to "sell" energy to neighboring countries, typically with negative prices. If this fails the grid becomes unstable, similar to a shortage of power.

The current power grid was not originally designed to handle increasing power demand, to reduce emissions or environmental impacts, to be energy efficient, or to integrate renewable energy sources [9]. A new electricity infrastructure needs to be created, which is able to improve management, monitoring, and use of electricity. Particularly, the new infrastructure needs to integrate new regenerative sources of energy without negatively affecting the performance of the power grid, be able to manage and regulate the intermittent power output of regenerative energy sources while keeping them at a constant level.

## 2.4 Smart Grid Architecture

The National Institute of Standards and Technology (NIST) defines the Smart Grid as a "... *modernization of the electricity delivery system so it monitors, protects and automatically optimizes the operation of its interconnected elements, from the central and distributed generator through the high-voltage transmission network and the distribution system, to industrial users and building automation systems, to energy storage installations and to end-use consumers and their thermostats, electric vehicles, appliances and other household devices".*[2]

In the Smart Grid approach, the current power grid will be enhanced by technology from the ICT world to enable a fine granular monitoring and control of power supply and demand. The ICT enhanced Smart Grid will allow system operators to analyze the status and behavior of the grid efficiently by providing real-time information about the grid. Distributed controls and diagnostic tools at transmission and distribution level will be able to reduce the occurrence of blackouts and disruptions by balancing power demand and supply [15]. The grid's capability to monitor the network status will be enhanced and components need to be installed that are able to dynamically reconfigure themselves. This will reduce the impact of power quality disturbances by enabling effective detection of and quick reaction to outages. Especially, a detailed smart metering of power consumption will be performed to enable the power supplier to know exactly by whom, when, and how the energy is being used and required, facilitating a better management of power supply, and therefore, a better quality of power.

To cope with the volatility of power production based on regenerative energy sources and varying power demand, the Smart Grid needs to establish new ICT-based information flows. On the one hand, power metering needs to be achieved that allows for a more fine granular prediction of power demand. On the other hand, communication infrastructures between power suppliers, producers, transporters, distributors, consumers, smart meters, and energy management systems need to be established. The enhanced information flow will additionally enable the reshaping of power demand, which is an important part of the Smart Grid solution. As the proportion of volatile power production increases, fluctuations in power production and demand cannot be covered by only adapting medium and peak power generation at the power supplier's side anymore. In addition, consumers of power are requested to partly shift their power demand according to current power availability. There are two scenarios, where power consumers may be involved in Smart Grid management: Either power consumers may be requested do reduce their demand during a shortage of power or they may be requested to increase their demand during a surplus of power (see

[2]Report to NIST in the Smart Grid Interoperability Standards Roadmap, 2009: www.nist.gov/smartgrid.

Section 2.3). Except these extreme scenarios, it would be a good idea to generally shift power demand to periods when renewable energy sources are highly available to use as much of this power production as possible and to reduce medium load power generation (see Section 2.2).

*Demand/Response (DR) management* achieves the automatic adaption of electricity demand of end-users (industry, companies, or households) based on the current electricity price and the state of the electricity grid [16]. DR management was developed around the turn of the millennium in the USA [17]. At that time, power outages occurred due to an overloading of the electricity grid. As a countermeasure, the customers of the power suppliers were obliged to immediately stop their power consumption when they got an emergency signal from the power supplier. DR mechanisms need to be extended and improved with respect to the Smart Grid. On the one hand, currently available DR mechanisms only consider situations concerning a shortage of power in the grid, a surplus of power in the grid is not yet explicitly addressed. On the other hand, a deeper integration of power customers needs to be achieved, as DR is currently focused on major power customers (e.g. industry). In the Smart Grid DR needs to be established in a broader scope. Especially, the integration of smart households needs to be evaluated (see Section 3.2), but also other approaches need to be investigated. Cold storage houses, e.g. are currently evaluated with respect to be included in DR management [18] as well as data centers [19].

# 3. SMART GRID SECURITY

This section discusses security challenges of the Smart Grid. First, Section 3.1 provides a general overview on security, including security goals and possible attackers in the Smart Grid scenario. Then, specific security challenges are discussed in detail. On the power consumer side, privacy is discussed in Section 3.2 as a major issue due to the creation of power consumption profiles. On the power supplier side, vulnerabilities of Smart Grid control systems are evaluated in Section 3.3. In both cases, security measures are presented that can be used to achieve specific security goals.

## 3.1 Security Overview

The Smart Grid interconnects ICT with power grid technology to improve its efficiency and reliability. However, the interconnection of different kinds of networks is a fundamental theoretical problem with far-reaching impacts. The interconnection results in a newly created *network of networks* with changed properties and novel phenomenons that need to be fully understood. Apart from being an interesting theoretical problem (e.g. in self-organization research), it is also highly relevant in practice and has serious real-world implications in the context of the Smart Grid.

Specifically, the interconnection of ICT and power grid networks leads to serious security threats and loopholes in the Smart Grid. The power grid, which was typically isolated in the past, gets directly or indirectly connected to public networks such as the Internet. Due to the complexity of this newly integrated network, it is a difficult task to prepare for the new challenges arising. The power grid becomes vulnerable to ICT-based attacks. On the one hand, privacy of user data is at stake. On the other hand, the Smart Grid is a critical infrastructure and security flaws may cause serious damage: Power production is threatened and, indirectly, the environment and the public health and safety is endangered (e.g. by massive blackouts). Therefore, the Smart Grid needs to be designed even more secure, reliable, and resilient than today's power grid to become broadly accepted by consumers.

### 3.1.1 Security Goals

This subsection lists technical security goals of networks [20] that need to be achieved within the Smart Grid:

- *Privacy and confidentiality:* In the Smart Grid, a smart metering infrastructure will be established to monitor power consumption at a fine granularity. The resulting power consumption profiles can be used to optimize power production and to maintain stability in the grid. However, such profiles can also be analyzed to gain and exploit further information. The power consumption of households can be analyzed as well as the power consumption of enterprises or administration. As power is consumed in every area of life, power consumption profiles cover nearly everything. Apart from power consumption profiles, further data is transmitted in the Smart Grid: Topology information (range, location) of the users infrastructure, device states, producer/supplier of devices, or data to identify and authenticate the power consumer is sent across the network. Data needs to be exchanged confidentially and, as far as possible, impersonalized between different roles in the market (as power production, transport, distribution, metering, or accounting). Privacy challenges of the Smart Grid are discussed in more detail in Section 3.2.
- *Data integrity:* Data integrity is important in different areas of the Smart Grid. The integrity of smart meter data, e.g. needs to be protected to

avoid energy stealing, where data may be forged to report low power consumption. Energy stealing was a problem even before the Smart Grid. An FBI study (2010) states that 10% of all smart meters in Puerto Rico are manipulated to report significantly reduced energy usage.[3] What is more, inaccurate information on power consumption may lead to instabilities in the grid if they are used for controlling power production. Also, in Smart Grid control systems (see Section 3.3), which control power production and distribution, data integrity is highly important. Modified or lost data may lead to malfunctions with possibly disastrous impact (e.g. in nuclear power plants).

- *Accountability:* Accountability is an issue with respect to power consumption as well as with respect to power production (photovoltaic, windmills). Electric cars, e.g., may be charged at different locations (supermarket or parking lot), but the energy bill is payed by the owner of the car.

- *Availability:* The availability of the Smart Grid is in question as the combination of energy network and ICT components creates new vulnerabilities to the highly critical power supply. Attackers with access to the Smart Grid communication system can institute failures or blackouts, which not only result in increased cost for both consumers and suppliers, but can even threaten national security since a stable power supply system is vital to society [21]. Another problem is the mutual dependency between ICT and grid technology in the Smart Grid infrastructure: Failure in ICT may lead to a failure in the power grid and vice versa (as ICT equipment is powered by energy). Furthermore, chain reactions are possible, where small blackouts may lead to wide-ranging blackouts.

- *Controlled access:* The Smart Grid is a large distributed system that is composed of millions of devices as control systems, servers, databases, workstations, management systems, smart meters, smart appliances, or gateways. Some of those devices are directly or indirectly connected to the Internet, which makes them accessible by attackers. Furthermore, each and every device that is connected to the Smart Grid is a potential entry point for intruders. Therefore, controlled access is highly important to restrict the usage of critical functions to authorized persons only.

---

[3]KrebsonSecurity:
http://krebsonsecurity.com/2012/04/fbi-smart-meter-hacks-likely-to-spread/.

### 3.1.2 Possible Attackers

The range of imaginable attacks is high and different groups may be interested in attacking the Smart Grid [1].

*Cyber terrorists* may use the Smart Grid to perform assaults. Denial of Service (DoS) attacks (e.g. by using botnets) may be performed to achieve directed local blackouts. Terrorists may try to interrupt the power supply of critical infrastructures as hospitals, water or gas supply, or of important industry. If they gain access to control systems, they may be able to initiate emergency procedures, e.g., shut down equipment to enforce minor or major blackouts. Alternatively, terrorists may try to destroy important control elements (by overheating or flooding) to cause damage, e.g., in nuclear power plants. It is also possible to physically attack the Smart Grid, by destroying important pylons in order to achieve a blackout chain reaction.

*Customers* may be motivated to tamper with smart meters to modify energy bills, either to reduce cost (consumption) or to increase benefit (production). Energy theft by consumers has always been a problem and is still causing massive financial losses to power suppliers. Also, getting access to power consumption profiles of other customers may be a motivation, this is discussed extensively in Section 3.2.

*Organized crime* may be interested in blackmailing power suppliers or customers. Criminals may get access to Smart Grid control systems and threaten to damage power plants to blackmail money from power suppliers. Alternatively, they may try to perform DoS attacks (using botnets) and threaten to cut important infrastructure off power supply. They may as well develop a universally exploitable vulnerability of smart meters, which can be used to either manipulate or stop them completely. Power customers can be blackmailed if criminals get access to power consumption profiles that reveal unpleasant information.

*Employees and service providers* are insiders that may manipulate the Smart Grid by sending fraudulent signals to smart meters or other critical parts of the grid, leading to serious security risks. Employees will also be exposed to social engineering attacks, where intruders establish social contact to create trust, before using USB sticks, manipulating web-sites, or e-mailing attachments to infect critical hosts. An example of this was presented at the RSA conference 2009 [1], where social engineering motivated an insider to execute a malicious attachment. By doing so, an intruder gained access to maintenance services of a nuclear power plant. Additionally, phishing will become a problem, where adversaries try to steal credentials of employees.

## 3.2 Privacy Challenges in Smart Households

One of the key challenges and major obstacles in the widespread deployment of the Smart Grid is user privacy. The Smart Grid relies on the usage of smart meters for billing and power consumption feedback purposes (as discussed in Section 2.4). Smart meters record and send fine granular readings of power consumption to the power suppliers. The measured power consumption data is way more detailed than the monthly or yearly measures in today's power grid and allows for the generation of power consumption profiles. These profiles, however, can be used for deriving usage patterns of persons, households, companies, or industry [22].

Households will be fully integrated in the future Smart Grid, in their role as power consumer, power producer, and in their capability to shift power consumption in time (DR management, see Section 2.4). Figure 3 illustrates a vision of a smart household: The power consumption of the household will be measured by *smart meters*, which are able to generate fine granular power consumption records, in contrast to old power meters that only report a cumulative power consumption. On the one hand, a smart meter enables the user to identify power consuming devices. Energy wasting devices can be replaced and the user is able to change usage behavior to reduce energy consumption. On the other hand, the power supplier is able to create power consumption profiles of the household, which enables a fine granular prediction of power demand to optimize power production processes. Power consumers are either smart appliances or usual devices. A

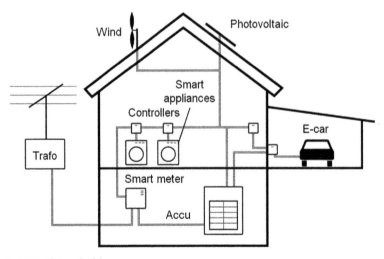

**Fig. 3.** Smart household.

smart appliance (e.g. smart washing machine, smart drying machine, smart air-conditioning, smart fridge) is controlled by the power supplier to reshape its energy consumption. Washing machines, e.g., are programmed by the user to do laundry up to a selected point of time. The power supplier is able to control the energy consumption by varying energy prices dynamically. The washing machine will start, e.g. if the price is lower than a certain threshold.

In addition to consuming energy, the household will be a producer of energy (e.g. by using windmills or photovoltaic) and the energy production will also be measured by smart meters. Furthermore, the smart house may be able to store energy in batteries. On the one hand, there may be an explicit accumulator, e.g. to store the energy produced by photovoltaic. On the other hand, an electric car may be available that is also equipped with powerful batteries that can be used as energy storage. Two usage scenarios of energy storage are possible: In the *customer-based* approach, the household powers its devices with energy from the storage/electric car during peak load consumption times, when energy is expensive. Later on batteries are recharged while energy is cheap (e.g. during the night). In the *supplier-based* approach available storage solutions are used by the power supplier similar to pumped-storage approaches. In the context of electric cars this is known as the *Vehicle to Grid* approach. Each of the cars has only little capacities, however, the sum of them have significant impact (the German government, e.g., plans to have 6 million electric cars in Germany by 2030).

### 3.2.1 Power Consumption Profiles

It is important to see that the integration of households into the Smart Grid management significantly increases the information flow between household and power supplier, opening up privacy and security challenges. Especially, the transmission of fine granular power consumption profiles is a problem, as such profiles can be analyzed down to a high level of detail. It is even possible to analyze households down to the fact which device they were using at what point of time or which TV program they were watching [23]. Figure 4 illustrates a (fictive) example of a household's power consumption profile. It can be observed that the power consumption of certain devices can easily be determined, as the fridge, heating, or the frequent power fluctuations caused by the hotplate.

Table 1 lists some examples of interest groups that might want to have access to power consumption profiles of households. Energy consultants, e.g., may want to consult households with high-energy consumption or specific energy usage to advise measures to achieve more energy efficiency.

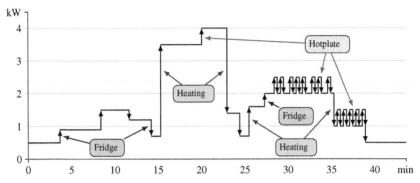

**Fig. 4.** Power consumption profile example.

**Table 1** Exploiting power consumption profiles of households [24].

| Interest Group | Purpose |
| --- | --- |
| Creditors | Determine behavior that indicates creditworthiness |
| Criminals | Identify times for burglary/high-priced appliances |
| Electricity advisory companies | Promote energy conservation and awareness |
| Employer | Investigate applicants or to monitor employees |
| Insurance companies | Determine health care premiums based on behavior |
| Landlords | Verify lease compliance |
| Law enforcers | Identify suspicious or illegal activity |
| Marketers | Profile customers for personalized advertisements |
| Private investigators | Monitor persons/specific events |
| The press | Get information about celebrities |

Insurance companies may try to determine health care premiums based on unusual behaviors that might indicate illness. Especially, it may be possible to derive a user's lifestyle from power consumption profiles: Proper cooking or junk food? Watching TV in the evening or going out everyday? Power suppliers could, e.g., sell power consumption profiles from individual households to marketers who use it to send personalized advertisements to the customers [25]. Law enforcers may use power consumption profiles to identify suspicious or illegal activity. The legal case Kyllo vs. USA (2001) is an example for this, where the government used utility bills to "*show that the suspect's power usage was "excessive" and thus "consistent with" a marijuana-growing operation*".[4] Landlords may use power consumption profiles to verify lease compliance (e.g., the number of persons that are living in an apartment). Private investigators would be able to monitor specific events, e.g., the time when somebody leaves the house, without physically supervising the person. Also, employers may be interested in power consumption profiles to investigate the lifestyle of applicants (similar to health care insurance companies). It may additionally be interesting to monitor employees, e.g., while they are on sick leave, to see if they are lying in bed or refurbishing the house. The press (paparazzi) may use power consumption profiles as a further source of information to report on the lifestyle and behavior of celebrities. Creditors will be interested in power consumption profiles to judge on the creditworthiness of persons by analyzing lifestyle and devices used in a household. Finally, criminals may use power consumption profiles to identify times for burglary (when nobody is home in the neighborhood), to identify alarm systems by their energy footprint, and to identify high-priced devices. Obviously, such aspects of consumer privacy are tightly linked to the problem of security [25].

Not only power consumption profiles of households are of interest, also the profiles of companies or industry may be exploited. Interesting questions that may be answered by power consumption profiles are: How is business going? Is there currently a high/low production volume? Is a company expanding? Do they have a new product (starting of new production line)?

### 3.2.2 Security Measures

Smart meters will be connected to a so-called *gateway*, which will be the central communication point between power supplier and power consumer. Several smart meters and smart appliances can be connected to the gateway which is able to control smart appliances. It is not yet quite clear, where such a gateway will be located. Location possibilities are, e.g. within the smart meter, as a separate device at the power consumer, or within the transformer that distributes power to consumers. The gateway may also act as energy management system for home automation, or even be a multi-utility gateway, responsible for water and gas metering. It is important to see that a gateway (which has more resources available than, e.g. smart meters) seems to be the natural point to enforce security and privacy policies. There are several privacy measures that could be performed by the gateway:

[4]Kyllo vs. USA: http://caselaw.lp.findlaw.com/scripts/getcase.pl?navby=CASE&court=US&vol=533&page=27.

- *Anonymization:* Power consumption data can be reported to the power supplier by using pseudonyms, as a concrete identity is not important for predicting future power consumption. Accounting information needs to be transmitted including identities of course, however, fine granular power consumption data is not needed for accounting and billing.

- *Temporal aggregation:* Power consumption data can be aggregated over time, e.g., only the overall power consumption within 1 h is transmitted, together with minimum and maximum consumption. This prevents a detailed analysis of usage behaviors.

- *Spatial aggregation:* Power consumption data of different customers can be aggregated (e.g. all customers that are provided by a single transformer). This prevents the analysis of a single customer. Accounting information needs to be transmitted separately in this case.

- *Charging of batteries:* It may be possible to power a household only by using batteries from an electric car or other available batteries. This way, only the battery charging is visible in the power consumption profile. Alternatively, it may be possible to control the charging of the batteries specifically to even out power consumption profiles to a certain extent, e.g., charge faster while no energy is consumed, and charge slower/stop charging while energy is consumed.

- *Homomorphic cryptography:* A very recent approach to solve the privacy problem in smart metered environments is the use of homomorphic encryption. This form of encryption allows several computations to be directly carried out on ciphertext instead of plaintext and still obtain a correct result in encrypted form. Examples for partly homomorphic encryption schemes are, e.g., ElGamal [26], Goldwasser-Micali [27], or Paillier [28]. However, the complexity of these schemes is currently much too high to use them in household applications. More details on homomorphic encryption schemes are given in Section 4.1.1.

In short, it becomes clear that it is of highest importance to keep power consumption data private and confidential as far as possible. This is most important, as power consumption data covers nearly every area of life. Generally, data need to be encrypted before transmitting it from the power consumer to the power supplier to achieve confidentiality during the transmission process. However, the operation of the Smart Grid should not be disturbed by any measures taken to ensure privacy or confidentiality [25].

## 3.3 Security Challenges in Smart Grid Control Systems

While security goals of Smart Grid control system networks generally do not differ from those in other networks, their importance is shifted in the context of the Smart Grid. Especially, availability is an important goal as the Smart Grid is a highly critical infrastructure, where outages have economical, ecological, and social consequences.

Power grid control systems are used to ensure a reliable and effective operation of the entire electric power system. They monitor and control power production and distribution processes, as e.g., opening and closing circuit breakers or setting thresholds for preventive shutdowns of power plants. Programmable Logic Controllers (PLC), such as SCADA networks, interconnect and control sensors and actors. Mostly they use proprietary protocols, e.g., the Process Fieldbus (PROFIBUS) [29]. ICT-based improvements of such systems (as discussed in Section 2.4) enable a wide-scale monitoring and controlling of Smart Grid components and promise to reduce time, money, and productivity losses caused by power quality fluctuations. However, the interconnection of ICT technologies with power grid technologies also leads to new security threats and loopholes in the integrated network. Power grid control systems used to be protected through isolation in the past and security was not an important issue in such systems. Instead, the focus was set on real-time features, functional safety, and performance of the control systems. In the Smart Grid, control systems achieve more and more communication capabilities and get directly or indirectly connected to public networks, making them vulnerable to ICT-based attacks. Attacks to the power grid, however, can cause serious damage as power production is threatened with possible social, ecological, and economical effects. The Stuxnet [30] worm (discovered in 2010) is an example for an attack to control systems as it targeted SCADA systems.

### 3.3.1 Vulnerabilities of Smart Grid Control Systems

The progress of a possible Smart Grid attack can be described similar to other network-based attacks:

1. acquisition of information about the target network infrastructure,
2. analysis of available security measures present, and
3. exploitation of vulnerabilities to mount an attack.

This subsection analyzes the first two steps of a possible attack with respect to power grid infrastructures. The third step, which results from the first two steps and would involve illegal actions, is only discussed. The focus is set on the assessment of some of the most severe security issues present in current Smart Grid control systems that can be exploited without deep security knowledge.

The first step to investigate control system vulnerabilities is the acquisition of information on the target network infrastructure. This includes acquiring information on both hard- and software by reviewing the manuals of the suppliers of these products. In this assessment, a web-based control system management interface, namely CAREL Pl@ntVisor and a remote control system management tool, Electro Industries/GaugeTech Communicator EXT, were investigated. The manuals of both products revealed information usable to search and access active applications on the web, such as detailed product names, product versions, default user names and passwords.

A simple Google search (September 2011) was performed and found altogether, 81 active control system management applications (61 Communicator EXT and 20 Pl@ntVisor). Even without providing any credentials, all of the found applications were accessible and revealed highly interesting information, as

- device's configuration settings,
- internal IP addresses and MAC addresses,
- authentication prompts revealing user names,
- error messages disclosing internal IP addresses,
- firmware and system versions.

The first approach in the investigation was to access the found active Pl@ntVisor applications, which can be remotely controlled by using a standard web-browser. In the tests, several of the applications were configured with default user names and passwords, in other cases the web search revealed valid user names (as depicted in Fig. 5) that could be used together with standard passwords, rendering those applications an easy target for attacks and opening up access to control systems.

A second approach to gain access to control systems was assessed by using the freely available management software Communicator EXT.[5] Communicator EXT can be used "...to custom configure Shark® meters, Nexus® 1500 m, Nexus® 1262/1272 m, Nexus® 1252 m, DM Series meters and Futura+ meters at local or remote sites and retrieve data from them for analysis". Using this software, it was possible to connect to all of the previously identified targets. Even more alarming was the fact that all connections to these systems could be

---

[5]Communicator EXT: http://www.electroind.com/pdf/ComEXT_Manual.pdf.

established using the vendor's default connection settings and the target device's IP address. The software revealed settings and statistics, allowed to download logs and settings, to reset device information and settings, and enabled the editing of advanced network settings as IP addresses, FTP servers, and protocols. Despite of the fact that some devices were protected by passwords, it was still possible to read information (as illustrated in Fig. 6).

In a third approach to gain access to control systems, known exploits of web-based applications were investigated. Pl@ntVisor is vulnerable to directory traversal attacks, where restricted directories of hosts can be accessed [31]. To exploit this vulnerability, URLs of targets need to be extended by expressions as " . . \ " to navigate though the server's file system. This effect can either be used to acquire the operating system's accounts and password hashes [32], or read Pl@ntVisor user names and passwords directly, as they are stored in plaintext format in a file. Figure 7 shows such a file containing the Pl@ntVisor credentials. Using this information, an adversary is able to log into the management system, read sensitive information, and perform manipulations.

In summary, gaining unauthorized access to grid control systems is possible by following simple procedures, without having deep security knowledge. Currently more than 90% of successful attacks take advantage of known

**Fig. 5.** Pl@ntVisor web search.

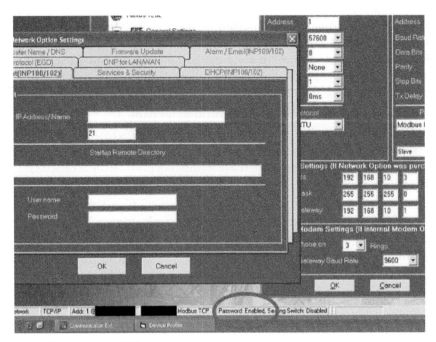

**Fig. 6.** Communicator EXT.

```
[Users]
Scheduler="797A7B".3433,31
PVRemote="",31,343731
Administrator="                    ",383131,343731
        ="            ",353B32,31
        ="            ",353B32,31

[HEADER]
VER=2
```

**Fig. 7.** User accounts and passwords.

vulnerabilities and misconfigured operating systems, servers, and network devices [33]. There are many more, often more complex, ways to gain access to a system. However, the assessed control system environments had not even implemented basic protection mechanisms, such as:

- *Firewalls:* In none of the investigated systems a firewall was present to restrict access to authorized staff.
- *Security configuration:* Critical systems were seriously misconfigured. Default passwords and default user names were used or authentication

was even turned off. Furthermore, some applications produced error messages that revealed sensitive information as user names.

- *Software management:* A periodical software patch process was not applied. The exploited software vulnerability was published in September 2011 [31]. In May 2012, the assessed systems still had not installed the required software patches or applied other counter measures.
- *Intrusion detection/prevention:* It can be assumed that there was no regular checking of access logs or active connections performed to detect anomalies and malicious behavior, as the assessed systems were still reachable in May 2012.

### 3.3.2 Security Measures

The vulnerabilities that were discussed in Section 3.3.1 can be faced by several security measures. As a first measure, a *firewall architecture* needs to be applied that is able to reduce the threat of being attacked at the network level. Figure 8 illustrates an example network of a power supplier including a firewall architecture. At the left part of the figure, the *corporate network* is illustrated that consists of publicly accessible servers (e.g. providing mail or web services), internal databases, and the staffs workstations. The corporate part of the network is typically accessible from the Internet. The *control system network* (depicted on the right side of Fig. 8) contains the control systems to monitor and control power production, transport, and distribution. The center part of Fig. 8 illustrates the *operation network*. This network is used by operators and administrators to monitor and manage devices in the control system network by using a Master Terminal Unit (MTU). Log- and data-acquisition servers are located in the operation network which is also connected to the corporate network for productivity reasons. An

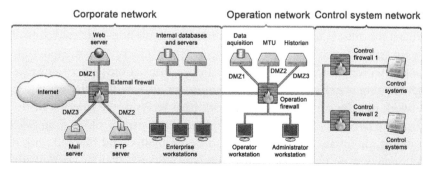

**Fig. 8.** Logical separation of networks by perimeter security devices.

*external firewall* needs to be applied to the corporate network to secure the external perimeter of the entire facility. It protects the corporate network against untrusted communication and creates demilitarized zones (DMZ) that separate hosts that need to be accessible from the outside from the rest of the network. An *operation firewall* separates the corporate network from the operation network and protects the operation network against any malicious connection coming from the corporate network. It creates a second set of DMZs to further separate services from each other, in case one of them gets compromised. Several *control firewalls* protect each of the control networks, separate them from each other, and allow communication with the corresponding MTUs only. The separation is able to prohibit an attacker that may have compromised a less important system to compromise other control systems.

In addition to installing firewalls, restrictive *firewall security policies* need to be defined. Any access or connection from and to control systems needs to be restricted, avoiding the use of generic rules that target many hosts or services simultaneously. Often, permissive firewall rules are generated that allow a wide range of IP addresses to access entire networks or a wide range of ports to support the communications flow and business continuity. This, however, opens up potential attack vectors as attackers gain access to a wide range of systems if they have compromised a host within the network. Therefore, access is granted only to those IP addresses and ports strictly necessary to perform required tasks, while any other access should be denied. It is also required to implement a per-user access control rule, as firewall rules that are solely based on IP addresses grant anyone who is using a certain host access to control systems. Also the privileges of control systems need to be restricted. The fact that control systems typically were physically isolated within facilities has created the misconception that control systems are trustworthy systems that can be granted unrestricted network access. However, if a control system gets compromised, it should not be able to contact the outside world or other critical systems. If a non-restricted control system gets infected, e.g., it is able to contact an attacker's command-and-control server in order to download further instructions and to enhance the ongoing attack. Therefore, it is important that neither the operation network nor the control system network have access to the Internet or the corporate network. Also, unneeded applications, as e-mail, file sharing, web browsing, or instant messaging need to be prohibited and blocked within the operation network and control system network. Besides configuring stringent rules, it is also important to have documentation that describes all allowed connections within the protected networks to prove the legitimation of connections. This includes, e.g., the source and destination of a connection, types of applications allowed to establish a connection (network protocol and port), and date and time when the establishment of a connection is allowed and how long the connection may remain established. Operators are able to periodically evaluate this information and use it as basis for detecting anomalies and malicious traffic to/from control systems.

Furthermore, it is also required to implement a sufficient *software management*. The installation of unauthorized software needs to be restricted to reduce the number of vulnerabilities in critical systems. A software patch management needs to be realized to ensure that patches, updates, and security fixes are installed consistently on a regular schedule. Figure 9 depicts a block diagram of a patch management process, based on a database containing the version numbers of software deployed in the power supplier's network. Additionally, the database is aware of available patches, updates, and disclosed vulnerabilities and exploits of deployed software. The patch management includes the testing and verification of patches as well as a backup strategy for the current system to be able to recover if problems occur. Patch management does not completely solve the problem of attackers exploiting a system's vulnerabilities, however, it reduces the risk of specific attacks significantly.

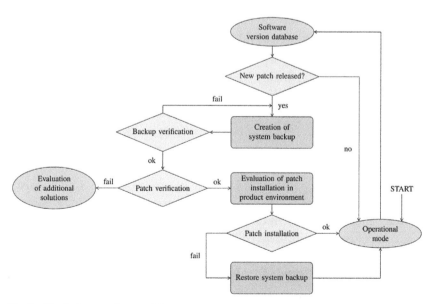

**Fig. 9.** Block diagram for a patch management process.

A further aspect to ensure the security and reliability of a system is the continuous monitoring and logging of all activities in the control system network to identify anomalies and fraudulent activities within the infrastructure. *Intrusion detection systems* (IDS) can be used to improve this process considerably. IDS are able to expand the visibility of security-related events and to improve the reaction capabilities of administrators as an IDS automates the monitoring process and analyzes events automatically. If IDS are added to the network architecture presented in Section 3.3.2, they should be placed between corporate and operation networks, and between operation and control system networks. This allows administrators to gain a complete overview of the activities between all network parts.

Although the discussed security measures help to significantly reduce vulnerabilities (as discussed in Section 3.3.1), these measures represent only a first step toward a secure Smart Grid. The focus of the suggested measures is the prevention of attacks that can be performed without deeper security knowledge. Advanced attackers, however, will have more possibilities to perform attacks to the Smart Grid. Also, the vulnerabilities discussed in Section 3.3.1 only represent a small part of the overall security challenges in the Smart Grid, such as social engineering, physical access to devices, or attacks from insiders, which need to be carefully investigated [1].

# 4. ENERGY EFFICIENCY VS. SECURITY

Energy efficiency and security are interrelated in many areas of the Smart Grid and both goals are often in conflict with each other. However, the two goals are equally highly relevant: On the one hand, the Smart Grid needs to improve energy efficiency and to integrate the volatile production of power based on renewable energy sources. On the other hand, the Smart Grid is a highly critical infrastructure that needs to be protected by security mechanisms. Therefore, energy efficiency needs to be carefully balanced against security measures. This section discusses the trade-off between energy efficiency and security in the context of two concrete examples. First, Section 4.1 analyzes the overhead caused by encryption of power consumption data of households. Second, Section 4.2 investigates obstacles in applying security mechanisms to Smart Grid control system networks.

## 4.1 Encryption of Power Consumption Profiles

Encryption of data is an important security service to preserve data confidentiality, inside as well as outside of the Smart Grid. Especially, power consumption data of costumers, billing information, and signaling messages need to be protected, as discussed in Section 3.2. The power consumption of data encryption in smart meters is used as an example for a highly relevant security service of the Smart Grid in this subsection.

Cryptography was suggested in Section 3.2.2 as security measure to achieve confidentiality. Considering the importance of having a secure Smart Grid implementation, the use of the most sophisticated cryptographic means available is indicated. However, the aspect of energy efficiency must not be forgotten while choosing the algorithms to employ. Not only economical aspects of saving energy have to be considered, also ecological impact is of great importance today. To achieve a balance between security and overhead in energy consumption, a brief analysis of energy efficiency and security is provided in [34], concerning different cryptographic algorithms. The following three algorithms are evaluated: The Advanced Encryption Standard (AES) [35] as an example of symmetric cryptography, the Rivest-Shamir-Adleman (RSA) algorithm [36] as an example of symmetric cryptography, and the Paillier cryptosystem (Paillier) as an example of homomorphic cryptography.

### 4.1.1 Cryptographic Algorithms

AES was developed in 1997, when the US National Institute of Standards and Technology decided to start an initiative to find a replacement for the Data Encryption Standard (DES). The ongoing evolution of computational power made this step necessary as the applied standards were no longer safe to use and could no longer provide adequate protection of data [37]. Several proposals were submitted and finally the Rijndael-Algorithm created by the Belgian scientists Joan Daemen and Vincent Rijmen was chosen because of its simplicity and high security standard [35]. AES is *symmetric cryptography* or secret key cryptography. Symmetric cryptography requires both parties, sender and receiver, to share a mutual secret key [38]. This key is used for encryption as well as for decryption of data. In general, symmetric cryptography is less complex and faster than other cryptographic methods but needs to deal with the problem of key sharing, meaning the transport of the secret key to the concerned parties without compromising it. Another problem of symmetric cryptography is scalability. For every new communication partner a new set of keys is needed. In a network with $n$ communication partners, this means $\frac{n \cdot (n-1)}{2}$ different keys are required [39]. The security of AES relies on the nonlinearity of its operations. A detailed explanation of AES is given in [37].

RSA was introduced in 1977 by Ronald Rivest, Adi Shamir and Leonard Adleman. It was the first asymmetric cryptosystem introduced to public [40]. In the late 1990s, it was revealed that Clifford Cocks from the British government created a similar algorithm already in 1975, however, it was highly classified and never disclosed to public. RSA is *asymmetric cryptography* or *public key cryptography*. In contrast to symmetric cryptography, asymmetric cryptography uses two different keys for encryption and decryption. One is called public key and is, as the name suggests, openly available and not kept secret. The second one is the private key, which needs to be kept secret and is only known to its owner. The idea of public key cryposystems was introduced in 1976 by Diffie and Hellmann [41]. The public and private key are mathematically connected and theoretically it is possible to derive the private key from the public key. This however is computationally infeasible with current technology and the security of asymmetric cryptography depends on this premise [39]. Asymmetric algorithms use more complex mathematical operations than symmetric algorithms, which derogates their performance and makes them more energy consuming [39]. The problem of key sharing, however, is less difficult as the public key is fully disclosed. Also, the number of keys needed is reduced from $\frac{n \cdot (n-1)}{2}$ to $n$ keys in a system with $n$ participants. Some public key systems can additionally be used for authentication (digital signatures), which is not possible with symmetric systems [41]. The mathematical background of RSA is discussed in detail in [39].

The Paillier cryptosystem is *homomorphic cryptography* and was proposed in 1999 by Pascal Paillier. In [42], homomorphic encryption schemes are described as "*encryption transformations mapping a set of operations on cleartext to another set of operations on ciphertext.*" Let $P$ be a mathematical group of plaintexts, $\oplus$ and $\otimes$ algebraic group operators and $E$ an encryption function. Formally homomorphism in cryptography is given as

$$\forall a,b \in P : E(a \oplus b) = E(a) \otimes E(b).$$

The notion of the existence of a *fully homomorphic* scheme was first proposed by Rivest, Adleman, and Dertouzos in [43] with the introduction of the RSA public encryption scheme in 1978. A fully homomorphic scheme in this context means a turing complete scheme with a combination of homomorphic operators with which every possible process can be executed on the cipher text without decrypting it. When decrypted, it shows the same result as the same operators executed on the plaintext [44]. Finding a fully homomorphic scheme has long been an important topic of cryptographic research. Only in 2009 Craig Gentry found the first fully homomorphic algorithm that supports addition as well as multiplication and is turing complete [45]. Many *partially homomorphic* schemes were found prior to this discovery. They only support a restricted number of operators. The first partially homomorphic scheme was discovered by accident by Rivest, Shamir, and Dertouzos in 1978 and is known to us as the asymmetric RSA algorithm. The fact that it is multiplicatively homomorphic was discovered shortly after its release and started the discussion about the possibility of fully homomorphic schemes [36]. A detailed analysis of the RSA homomorphic property is given in [46].

### 4.1.2 Energy Consumption of Encryption in the Smart Grid

The three discussed cryptographic systems were evaluated in terms of energy consumption and scalability with respect to their usability in the Smart Grid. The Smart Grid architecture as it is depicted in Fig. 10 was simulated to evaluate the different algorithms. Smart meters ($s_1, \ldots, s_n$) are installed in every household of the Smart Grid architecture. Each of the households is part of a cluster ($c_1, \ldots, c_m$) that bundle smart meters of a certain area. These clusters are illustrated by circles in Fig. 10. Each cluster is in possession of a data aggregator ($a_1, \ldots, a_m$). All smart meters in one cluster send their power consumption information to their respective data aggregator (e.g. a gateway in the transformer). The aggregator combines the information received from the smart meters to reduce message overhead and also to obfuscate the individual power consumption data. In this scenario, the power provider receives detailed information regarding the clusters only, not the power consumption profiles of individual households. The granularity of the aggregation needs to remain high enough to enable the power supplier to efficiently adjust power supply to power demands. For billing issues, e.g. a summary of each household's energy consumption per month is sufficient [47].

Five different scenarios (S1–S5) were assessed that vary in two parameters: The cluster size ($n$) and the overall number of clusters ($m$). The scenarios were designed in increasing complexity:

S1: $n = 50, m = 1$,
S2: $n = 1, m = 50$,
S3: $n = 1, m = 500$,
S4: $n = 500, m = 1$,
and S5: $n = 50, m = 50$.

These scenarios include extreme cases where only a single smart meter is connected to each aggregator (with an increasing number of aggregators) and more realistic cases with 50/500 smart meters per aggregator. Simulation steps included the encryption and sending of metering data from households

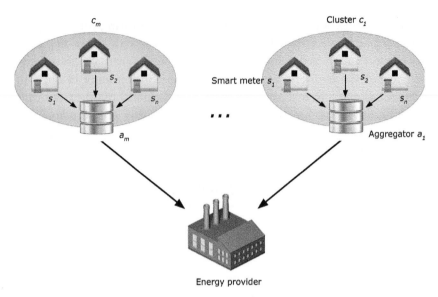

**Fig. 10.** Smart Grid architecture used for the simulations.

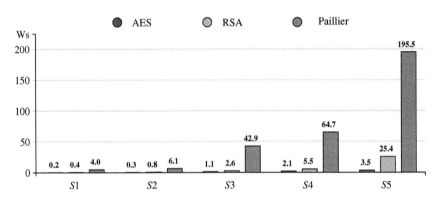

**Fig. 11.** Average energy consumption.

in the Smart Grid to the power supplier (via aggregators) who decrypts the received information. In Fig. 11 the average energy consumption is compared according to all scenarios and encryption methods.

It can be observed that AES uses the least complex operations, which results in being the most energy efficient and resource-saving method. Even in the largest scenario, the values for duration and energy consumption do not increase significantly and remain steady at a relatively low level. In contrast to the high energy efficiency of the AES algorithm, the symmetric encryption method in general faces the problem of a missing key sharing. A possible, yet not recommendable, solution would be to hard code necessary keys into Smart Grid devices. RSA, as a representative algorithm of asymmetric cryptography, has a higher energy consumption and resource load than AES. While similar to symmetric encryption in the smaller scenarios, the duration and energy consumption increases about 5–7 times faster than symmetric encryption in the large scenario. The problem of key distribution is not as significant as with symmetric encryption, but still present: Communication with key authorities that manage the private and public key authentication and dispensation needs to be considered. The Paillier algorithm (and homomorphic cryptography as a whole) uses highly complex mathematical computations and consumes most energy and resources. Energy consumption is several times higher than that of the other encryption methods. Therefore, homomorphic cryptography is currently insufficient for an environment that desires to save energy rather than consume it [48]. The highly important advantage of this algorithm, however, is the achieved privacy. With homomorphic encryption, user data does not have to be decrypted while processing it (e.g. for aggregation). This means that no part other than the smart meter itself is aware of the detailed energy consumption information of a single household. This kind of privacy cannot be achieved by symmetric or asymmetric encryption.

Additionally, an extrapolation to a Smart Grid with 40 million households was calculated (Germany had about 40 million households in 2012[6]). This Smart Grid consists of 1 million clusters with 40 smart meters per cluster. Table 2 illustrates the power consumption of encryption and the energy consumption during 24 h, if a households's energy consumption data is sent every minute. Table 2 may be used as a basic guideline on the applicability of different encryption mechanisms in the Smart Grid. It can be observed that asymmetric encryption needs about eight times more energy than symmetric encryption and the energy consumption of homomorphic encryption is about eight times higher than that of asymmetric encryption.

The Paillier cryptographic system offers the possibility to aggregate user data without the need to decrypt it. With only a single encryption/decryption process during the whole data transfer, several operations can be saved compared to asymmetric and symmetric encryption, possibly leading to energy savings. To evaluate this, additionally a hierarchical aggre-

---

[6]Number of households in Germany 2012: http://de.statista.com/statistik/daten/studie/1240/umfrage/.

**Table 2** Extrapolation of the energy consumption of a Smart Grid to 40 million households.

|          | Average Consumption (W s) | 24 h        |
|----------|---------------------------|-------------|
| **AES**     | 48,160                    | 19.3 kW h   |
| **RSA**     | 407,200                   | 162.9 kW h  |
| **Paillier** | 312,7200                  | 1250.9 kW h |

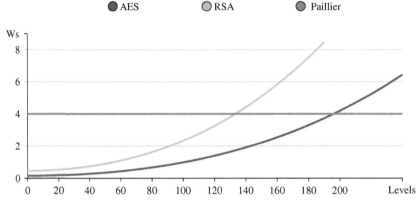

**Fig. 12.** Number of hierarchical aggregator levels and respective energy consumption of encryption and decryption.

gation architecture was analyzed, where several cascaded aggregator levels were used between the smart meter and the power provider. It was computed, how many aggregator levels would be needed to reach a break–even point, where homomorphic aggregation consumes less energy than other cryptographic systems. The result for a scenario based on $S1$, meaning one cluster handles 50 smart meters, is depicted in Fig. 12. As expected, the energy consumption of encryption and decryption with respect the Paillier cryptosystem is independent of the number of aggregation levels. However, it becomes apparent that the number of aggregator levels that would be required to even out the energy consumption of the Paillier cryptosystem, is unreasonably high: 191 aggregators for symmetric and 135 aggregators for asymmetric encryption would be needed, both numbers are far too high for realistic implementations.

As this brief analysis shows, none of the aforementioned encryption methods is the ideal solution for smart meters in the Smart Grid. All methods have significant advantages and disadvantages that make them more or less suitable but no perfect solution can be found to balance energy efficiency and security. A combination of the discussed methods needs to be used to achieve satisfactory efficiency and security. For the encryption of user power consumption data, a symmetric algorithm as AES is most appropriate, as it is a widely available and energy efficient solution. For key distribution an asymmetric algorithm, as RSA, can be adopted. This approach reduces the risk of key compromising and offers the possibility to adapt dynamically to new smart meters coming into the system. Although homomorphic encryption has several benefits for the Smart Grid, the research in this field is still not mature enough. Despite there are positive trends, available algorithms are still too resource consuming to be considered for use in the Smart Grid.

The goals of efficiency and security are not only conflicting in the field of user privacy in smart households but in many areas of the Smart Grid architecture (as it is described in Section 2.4). The following subsection describes conflicts that are raised at the power supplier side in Smart Grid control systems (see Section 3.3).

## 4.2 Obstacles in Applying Security Measures to Smart Grid Control Systems

Section 3.3.2 discusses a set of security measures that provide basic security against some of the most obvious and easy to exploit vulnerabilities of Smart Grid control system networks. However, although the suggested set of security measures represents commonly used and widely applied technologies of the ICT world, there are some major issues that may prevent their application within the Smart Grid. This subsection discusses several examples of widely used security measures that may be challenging within the Smart Grid:

- *Firewalls:* The missing separation of different parts of the network can cause major loopholes, as direct paths to control systems can be established. Firewall architectures can be applied to restrict access to control systems exclusively to authorized entities. In power grid control systems, however, firewalls are often not used due to a possibly negative influence on network performance [1]. As control systems often require real-time traffic, intermediate firewalls can negatively influence their operation by delaying network communication. This results in increased delays and increased reaction time to certain events, leading to a decreased overall efficiency in the grid, disturbances, or even to instabilities. Another issue with firewalls in power grid control systems is that they need to be configured to the highly specialized protocols of such systems. Furthermore, the firewall policies need to be updated with every change in the control system network, causing significant overhead.

- *Security configuration:* As control systems and operation network may be accessible via the Internet, they need to have carefully selected configuration settings. This includes the usage of measures like secured communication protocols and strong authentication. While such a task seems easily achievable, it is a perfect example of contradicting goals in the Smart Grid. Often authentication is not used in control systems due to the possibility of the so-called *lock-out* effect [1], where quick access to a system cannot be achieved. In the power grid immediate interaction with a control system is required during emergency situations. In the case of a lock-out, this time-critical interaction with control systems is prevented, e.g. due to a forgotten password, a missing smartcard, or a malfunction of a biometric system. The lock-out effect is the main reason why techniques such as biometric systems, smartcards, or strong passwords are often not applied in the power grid.
- *Intrusion detection/prevention:* As Smart Grid control systems are highly critical, intrusions and security breaches need to be detected as fast as possible. Hence, the usage of a power supplier's network needs to be carefully monitored and logged by IDS, or more advanced, intrusion prevention systems (IPS) [49]. Especially the use of IPS that are able to actively respond to attacks, however, is highly challenging in Smart Grid control systems. If a control system protocol or activity is falsely interpreted as an intrusion, an IPS active response to this process could interrupt or shutdown a critical process, leading to economically and ecologically costly malfunctions of the Smart Grid.

In short, these simple examples show that the efficient and effective operation of the Smart Grid is often affected by security measures that are themselves highly relevant to secure the operation of the Smart Grid.

## 5. RELATED WORK

The areas discussed in this chapter include the Smart Grid with respect to energy efficiency and possible security implications. While both topics are well researched in separation, finding a trade-off between the contradicting goals is still an unsolved challenge. Therefore, this section covers the research approaches in the fields of Smart Grid, security, and energy efficiency as well as their interrelation.

Smart Grid research and related research fields are currently very active, especially since the mandatory implementation of the system was decided by governments in the USA and wide parts of Europe in 2010. The Smart

Grid is thereby often described with properties as self-healing, highly reliable, optimized energy management, resilient to cyber attacks, and real-time pricing [50]. Further general information on Smart Grid technologies and arising challenges can be found in [51–53].

In the area of Smart Grid security related research, most of the current topics either cover privacy or security in smart meter environments. The potential privacy impacts of smart metering are, e.g. analyzed by [54, 55], concluding that privacy enhancing technologies, as anonymization or data aggregation have to be applied to preserve users' private data. While the resistance to attacks should be one of the key characteristics of the Smart Grid, this goal is hardly achievable in a real-life implementation of a Smart Grid due to several reasons, which are addressed in [56, 57], or [58]. The Smart Grid increases the complexity of existing power grid systems, as it will merge the power distribution system with an ICT communication infrastructure. While the increased complexity makes the overall system more vulnerable to attacks, the sheer size of a fully developed Smart Grid, several millions of nodes, imposes an additional challenge to security. The topic of critical control/SCADA systems in particular is analyzed with focus on security issues in [59, 60], or [61].

The relation of the contradicting goals of security and energy efficiency is an important topic in various fields. A generic analysis of an energy-security trade-off is given in [62, 63]. Similar research was done in the context of cipher algorithms and security protocols, which are analyzed in [64, 65]. The combination of the topics energy-efficiency and security has become especially relevant during the development of recent, highly decentralized, and resource-constrained technologies, such as wireless sensor networks. Here, limited energy reserves have to be carefully balanced with the security of the network. This is highly important due to the dissolving security perimeter in these scenarios. The impact of security mechanisms on the energy consumption in wireless sensor network scenarios is analyzed in [66].

# 6. CONCLUSION

This chapter describes the current situation of the power grid and its evolution to the Smart Grid. Challenges, as the integration of renewable energy sources or the reshaping of power demand, are analyzed together with management approaches to face such challenges. Furthermore, this chapter outlines problems hat arise through the interconnection of the power grid with information and communication technology. This interconnection leads to new privacy and security issues that need to be solved in the future Smart Grid.

Two major problems are discussed in detail: privacy challenges in smart households and security challenges in Smart Grid control systems. It becomes clear that the Smart Grid is an environment where a balance needs to be found between achieved energy efficiency and security risks imposed. The main goal of the Smart Grid is its efficiency and the successful integration of renewable energy sources. However, as the Smart Grid is a highly important infrastructure, where unavailability leads to social, ecological, and economical damage, the security of the Smart Grid plays a major role.

In future work, the complexity arising from the interconnection of power grid and information and communication technology needs to be further analyzed, especially with respect to security challenges. This is highly important, as the power grid is a critical infrastructure that needs to be carefully protected against malicious adversaries.

## ACKNOWLEDGMENTS

This work has been partly supported by the EC's FP7 All4green project (Grant No. 288674), by the EC's FP7 Network of Excellence EINS (Grant No. 288021), and by "Regionale Wettbewerbsfähigkeit und Beschäftigung," Bayern, 2007–2013 (EFRE) as part of the SECBIT project (http://www. secbit.de).

 ## LIST OF ABBREVIATIONS

| AES | Advanced Encryption Standard |
| DMZ | Demilitarized Zone |
| DoS | Denial of Service |
| DR | Demand-Response |
| ICT | Information and Communication Technology |
| IDS | Intrusion Detection Systems |
| IPS | Intrusion Prevention Systems |
| MTU | Master Terminal Unit |
| NIST | National Institute of Standards and Technology |
| Pailler | Paillier Cryptosystem (homomorphic cryptography) |
| RSA | Rivest-Shamir-Adleman Algorithm (asymmetric cryptography) |
| SCADA | Supervisory Control and Data Acquisition |

# REFERENCES

[1] C. Eckert, Sicherheit im Smart Grid – Eckpunkte für ein Energieinformationsnetz, Tech. Rep., Alcatel-Lucent Stiftung, 2011.

[2] The Economist, Building the Smart Grid, The Economist Newspaper Limited, 2009. <http://www.economist.com/node/13725843>.

[3] M. Weatherford, North American Electric Reliability Corp., Unknown unknows and the electric grid, SC Magazine.

[4] L.D. Kannberg, D.P. Chassin, J.G. De Steese, S.G. Hauser, M.C. Kintner-Meyer, R.G. Pratt, L.A. Schienbein, W.M. Warwick, GridWise TM: The Benefits of a Transformed Energy System: The Benefits of a Transformed Energy System, Tech. Rep. nlin.AO/0409035, Pacific Northwest Nat. Lab., Richlands, VA, September 2004.

[5] A. Abel, Smart Grid Provisions in H.R. 6, 110th Congress, Tech. Rep., Congressional Research Service (CRS), December 2007.

[6] Accent Energy, NY, The American power grid and electricity, 2012. <http://www.accentenergy.com/Energy101/ElectricityArticles.aspx/21_The-American-Power-Grid-and-Electricity>.

[7] Litos Strategic Communication, The Smart Grid: An Introduction, Tech. Rep., US Department of Energy, 2008, pp. 7, 14–19, 22.

[8] Federal Communications Commission, The National Broadband Plan, Chapter 12: Energy and the Environment, 2010, pp. 249–251.

[9] A. Battaglini, J. Lilliestam, C. Bals, A. Haas, The supersmart grid, in: European Climate Forum, Potsdam Institute for Climate Impact Research, 2008.

[10] A. Brautsch, B. Goll, R. Hestermann, T. Peter, M. Rieck, L. Timmermann, Leistungsreserve zur Absicherung von erneuerbaren Energien, Energieerzeugung KOMPAKT, EW (1) (2011) 8–11.

[11] GlobalData, Grid Integration of Renewable Energy Resources – Issues and Solutions, Tech. Rep., March 2011.

[12] B. Kennedy, Power Quality Primer, McGraw-Hill Professional, 2000.

[13] E.P.R. Institute, Estimating the cost and benefits of the Smart Grid, March 2011. <http://my.epri.com/portal/server.pt?Abstract_id=000000000001022519>.

[14] T. Baumeister, Literature Review on Smart Grid Cyber Security, Tech. Rep., University of Hawaii, Honolulu, 2010.

[15] US Department of Energy, Smart Grid System Report, Smart Grid System Report.pdf, July 2009. <http://energy.gov/sites/prod/files/2009>.

[16] E. Koch, M. Piette, Architecture Concepts and Technical Issues for an Open, Interoperable Automated Demand Response Infrastructure, Tech. Rep., Ernest Orlando Lawrence Berkeley National Laboratory, Berkeley, CA, USA, 2007.

[17] C. Brönniman, Demand Response – Eine neue Herausforderung für LonMark, December 2008. <http://www.lonmark.ch/PDF/fachberichte/demand-response.pdf>.

[18] A. Becker, U. Arndt, J. Hermsmeier, Flexibilisierung der Stromnachfrage, 2012. <http://www.energy20.net>.

[19] S. Klingert, A. Berl, M. Beck, R. Serban, M. Di Girolamo, G. Giuliani, H. De Meer, A. Salden, Sustainable energy management in data centres through collaboration, in: Proceedings of the First International Workshop on Energy-Efficient Data Centres (E2DC12), Lecture Notes in Computer Science (LNCS), vol. NA, Springer Verlag, 2012, p. NA.

[20] G. Schaefer, Security in Fixed and Wireless Networks, second ed., John Wiley & Sons, Ltd., 2003.

[21] H. Khurana, M. Hadley, N. Lu, D.A. Frincke, Smart-grid security issues, IEEE Security and Privacy 8 (1) (2010) 81–85.

[22] S. Iyer, Cyber security for Smart Grid, cryptography, and privacy, International Journal of Digital Multimedia Broadcasting (2011). <http://dblp.uni-trier.de/db/journals/ijdmbc/ijdmbc2011.html#Iyer11>.

[23] Heise Online, Smart Meter verraten Fernsehprogramm, March 2012. <http://www.heise.de/security/meldung/Smart-Meter-verraten-Fernsehprogramm-1346166.html>.

[24] The Smart Grid Interoperability Panel – Cyber Security Working Group, Potential Privacy Impacts that Arise from the Collection and Use of Smart Grid Data, Tech. Rep., National Institute of Standards and Technology, 2010.

[25] C. Wolf, A. Cavoukian, J. Polonetsky, Smartprivacy for the Smart Grid: Embedding Privacy into the Design of Electricity Conservation, Tech. Rep., Information and Privacy Commissioner (IPC), 2009.

[26] T. Elgamal, A public key cryptosystem and a signature scheme based on discrete logarithms, IEEE Transactions on Information Theory 31 (4) (1985) 469–472, http://doi.acm.org/10.1109/TIT.1985.1057074.

[27] S. Goldwasser, S. Micali, Probabilistic encryption and how to play mental poker keeping secret all partial information, in: Proceedings of the 14th Annual ACM Symposium on Theory of Computing, STOC'82, ACM, New York, NY, USA, 1982, pp. 365–377, http://doi.acm.org/10.1145/800070.802212.

[28] P. Paillier, Public-key cryptosystems based on composite degree residuosity classes, in: J. Stern (Ed.), Advances in Cryptology – EUROCRYPT'99, Lecture Notes in Computer Science, vol. 1592, Springer, Berlin/Heidelberg, 1999, pp. 223–238, <http://dx.doi.org/10.1007/3-540-48910-X_16>.

[29] K. Bender, Profibus: The Fieldbus for Industrial Automation, Prentice-Hall, Inc., 1993.

[30] N. Falliere, L.O. Murchu, E. Chien, W32.Stuxnet Dossier, Tech. Rep., Symantec, 2011.

[31] L. Auriemma, SCADA Advisories, SCADA security vulnerabilities, 2012. <http://aluigi.altervista.org/adv.htm>.

[32] J. Visser, On NT Password Security, May 1997. <http://www.mgforum.blair.at/doc/ntpass.pdf>.

[33] S.M. Amin, Smart Grid: overview, issues and opportunities: advances and challenges in sensing, modeling, simulation, optimization and control, in: Semi-Plenary Talk at the 50th IEEE Conference on Decision and Control (CDC) and European Control Conference (ECC), IEEE, Orlando, Florida, 2011.

[34] M. Zirm, Performance Comparison of Cryptographic Algorithms in Smart Grid Applications, Bachelor's Thesis, March 2012.

[35] National Institute of Standards and Technology, Security Requirements for Cryptographic Modules. Security Specifications for Cryptographic Modules Utilized within Security Systems Protecting Sensitive Information in Computer and Telecommunication Systems, Tech. Rep., 2001.

[36] R.L. Rivest, A. Shamir, L. Adleman, A method for obtaining digital signatures and public-key cryptosystems, Commun. ACM 21 (2) (1978) 120–126, http://doi.acm.org/10.1145/359340.359342.

[37] V. Rijmen, J. Daemen, The Design of Rijndael: AES – The Advanced Encryption Standard, Springer Verlag, Berlin, Heidelberg, 2002.

[38] J. Benoit, An Introduction to Cryptography as Applied to the Smart Grid, Tech. Rep., Cooper Power Systems, 2011.

[39] J. Swoboda, S. Spitz, M. Pramateftakis, Kryptographie und IT-Sicherheit, Vieweg + Teubner Verlag, 2011.

[40] H. Delfs, Introduction to Cryptography: Principles and Applications, Springer, 2007.

[41] D. Waetjen, Kryptographie, Spektrum Akademischer Verlag, Heidelberg, 2008.

[42] J. Domingo-Ferrer, A provably secure additive and multiplicative privacy homomorphism, in: Lecture Notes in Computer Science, vol. 2433/2002, 2002, pp. 471–483.

[43]  R. Rivest, L. Adleman, M. Dertouzos, On data banks and privacy homomorphism, Foundations of Secure Computation.

[44]  K. Henry, The Theory and Applications of Homomorphic Cryptography, Master's Thesis, University of Waterloo, 2008.

[45]  C. Gentry, A Fully Homomorphic Encryption Scheme, Master's Thesis, Stanford University, 2001.

[46]  K. Hayat, R. Brouzet, N. Islam, W. Puech, Analysis of homomorphic properties of RSA-based cryptosystem for image sharing, in: IEEE 10th International Conference on Signal Processing (ICSP), 2010.

[47]  G. Kalogridis, C. Efthymiou, Smart Grid privacy via anonymization of smart metering data, in: First IEEE International Conference on Smart Grid Communications (Smart-GridComm), 2010.

[48]  N. Lu, D.A. Frincke, H. Khurana, M. Hadley, Smart-grid security issues, IEEE Security and Privacy 8 (2010) 81–85.

[49]  K. Scarfone, P. Mell, Guide to intrusion detection and prevention systems (IDPS), NIST Special Publication 800-94, 2007.

[50]  R.E. Brown, Impact of Smart Grid on distribution system design, in: Power and Energy Society General Meeting – Conversion and Delivery of Electrical Energy in the 21st Century, IEEE, 2008, pp. 1–4.

[51]  S.M. Amin, B.F. Wollenberg, Toward a Smart Grid: power delivery for the 21st century, IEEE Power and Energy Magazine 3 (5) (2005) 34–41.

[52]  H. Farhangi, The path of the Smart Grid, IEEE Power and Energy Magazine 8 (1) (2010) 18–28.

[53]  J. En-Bo, Smart Meter System Design in Smart Grid Advanced Metering Infrastructure AMI, Tech. Rep., Electrical Measurement & Instrumentation, 2010.

[54]  E.L. Quinn, Smart metering and privacy: existing laws and competing policies, Social Science Research Network. <http://ssrn.com/paper=1462285>.

[55]  A. Cavoukian, J. Polonetsky, C. Wolf, SmartPrivacy for the Smart Grid: embedding privacy into the design of electricity conservation, Whitepaper, 2009. <http://www.futureofprivacy.org>.

[56]  A.R. Metke, R.L. Ekl, Smart Grid Security Technology, Tech. Rep., Motorola, Inc., 2010.

[57]  W.F. Boyer, S.A. McBride, Study of Security Attributes of Smart Grid Systems – Current Cyber Security Issues, Tech. Rep., Idaho National Laboratory, Critical Infrastructure Protection/Resilience Center, 2009.

[58]  A. Lee, T. Brewer, Smart Grid Cyber Security Strategy and Requirements, Tech. Rep., The Cyber Security Coordination Task Group, Advanced Security Acceleration Project – Smart Grid, 2009.

[59]  M.R. Chaffin, S.M. Tom, D.G. Kuipers, W. Boyer, Common Cyber Security Vulnerabilities Observed in Control System Assessments by the INL NSTB Program, Tech. Rep., INL Report to the Department of Energy, INL/EXT-08-13979, 2008.

[60]  M. McQueen, W. Boyer, T. McQueen, S. McBride, Empirical estimates of 0 day vulnerabilities in control systems, in: Proceedings of the SCADA Security Scientific Symposium 2009 (S4), 2009, pp. 6-1–6-26.

[61]  K. Barnes, National SCADA Test Bed Substation Automation Evaluation Report, Tech. Rep., INL Report to the Department of Energy, INL/EXT-09-15321, 2009.

[62]  S. Jahr, Security Versus Power Consumption, Master's Thesis, Gjøvik University College Department of Computer Science and Media Technology, 2006.

[63]  N.R. Potlapally, S. Ravi, A. Raghunathan, N.K. Jha, Analyzing the energy consumption of security protocols, in: Proceedings of ISLPED'03, Seoul, Korea, 2003.

[64]  N.R. Potlapally, S. Ravi, A. Raghunathan, N.K. Jha, A study of the energy consumption characteristics of cryptographic algorithms and security protocols, IEEE Transactions on Mobile Computing 5 (2) (2006) 128–143.

[65] L. Batina, J. Lano, N. Mentens, S.B. örs, B. Preneel, I. Verbauwhede, Energy, performance, area versus security trade-offs for stream ciphers, in: The State of the Art of Stream Ciphers, Workshop Record (2004), ECRYPT, 2004, pp. 302–310.
[66] C.-C. Chang, S. Muftic, D. Nagel, Measurement of energy costs of security in wireless sensor nodes, in: Proceedings of 16th International Conference on Computer Communications and Networks 2007, ICCCN 2007, 2007, pp. 95–102 10.1109/ICCCN.2007.4317803.

## ABOUT THE AUTHORS

**Andreas Berl** obtained his Ph.D. at the University of Passau (Germany) in 2011. He is currently working as researcher in the Computer Networks and Communications group at the University of Passau, chaired by Prof. Hermann de Meer. His research interests include energy efficiency, virtualization, and peer-to-peer overlays. Currently he is involved in the BMBF project "G-Lab_Ener-G — Improving the Sustainability of G-Lab Through Increased Energy Efficiency" and in the EU project "All4Green — Active collaboration in data centre ecosystem to reduce energy consumption and GHG emissions" (STREP, FP7). He is member of the EU Networks of Excellence "EuroNGI/EuroFGI/EuroNF — Design and Engineering of the Next Generation Internet" and "EINS - Network of Excellence in Internet Science" and the COST Action IC0804 "Energy Efficiency in Large Scale Distributed Systems". In 2009 he had a DAAD scholarship at Lancaster University, UK, supervised by Prof. David Hutchison.

**Michael Niedermeier** received his Diploma in Computer Science in 2009 from the University of Passau. Since then, he is working as a research associate at the Chair of Computer Networks and Computer Communications and at the Institute of IT Security and Security Law (ISL) at the University of Passau. His main research areas focus on energy efficient security concepts, security and functional safety in distributed systems like sensor networks or the Smart Grid. Currently, he is working on the EFRE-funded SECBIT project, whose goal is to support SMEs to strengthen their IT security and safety awareness. Additionally, he is a member of the EU-funded network of excellence "EINS", which offers a platform for worldwide cooperation and interdisciplinary research of the Future Internet.

**Hermann de Meer** is currently appointed as Full Professor of computer science (Chair of Computer Networks and Communications) and is director of the Institute of IT Security and Security Law (ISL) at the University of Passau. He had been an Assistant Professor at Hamburg University, Germany, a Visiting Professor at Columbia University in New York City, USA, Visiting Professor at Karlstad University, Sweden, a Reader at University College London, UK, and a research fellow of Deutsche Forschungsgemeinschaft (DFG). He chaired one of the prime events in the area of Quality of Service in the Internet, the 13th international workshop on quality of service (IWQoS 2005, Passau). He has also chaired the first international workshop on self-organizing systems (IWSOS 2006, Passau) and the first international conference on energy-efficient computing and networking (e-Energy 2010, Passau). He currently holds several research grants funded by the Deutsche Forschungsgemeinschaft (DFG) and by the EU (FP6 and FP7).

**CHAPTER FIVE**

# Energy Efficiency Optimization of Application Software

## Kay Grosskop and Joost Visser
Software Improvement Group (SIG), Amstelplein 1, 1096 HA Amsterdam, Netherlands

## Contents

*Advances in Computers*, Volume 88
ISSN 0065-2458, http://dx.doi.org/10.1016/B978-0-12-407725-6.00005-8

## Abstract

The application software design has a major impact on the energy efficiency of a computing system. But research on the subject is still in its infancy. What is the energy efficiency of software? How can it be measured? What are guidelines for the development of energy efficient software? In this paper, we set out to find an answer to these questions and motivate the need for a dedicated research area for application software energy efficiency. We place this subject in the context of other initiatives for green and sustainable computing and clarify central concepts. Furthermore we give an overview of different existing approaches for both measuring and optimization of software energy efficiency.

# 1. INTRODUCTION

IT as a substantial consumer of energy has become widely acknowledged nowadays. However, and for a good reason, most of the effort in analyzing and improving the energy efficiency of IT systems has been at the places where the vast amount of electricity is actually consumed: the computer hardware and the data center infrastructure. In this article we want to propose a shift in perspective to what we think is an essential key to the design of energy efficient systems that still has received little attention from academics and practitioners: the high-level application software.

There are two aspects to this. In one sense we are taking the application as the primary unit of evaluation and design for energy efficiency because this is where crucial decisions for computing systems are taken, be it from a user, designer, operator, or an investor point of view. Second, we want to raise the level of analysis from the hardware, data center infrastructure, or low-level system software to the high-level application software. The application level is crucial for the development of energy efficient systems because this is where fundamental decisions on system architecture are taken and where detailed knowledge about requirements is available.

Other areas of research on energy efficient system design are much more evolved for various reasons: for example, circuit architecture and hardware have been suffering from problems of heat dissipation, mobile and embedded devices struggle with a limited energy budget, and data centers infrastructure suffers from limitations of energy provisioning and heat dissipation as well as high energy bills. Moreover, because energy is dissipated in hardware, a natural inclination is to focus first on hardware optimization or at least the low-level system software that directly controls hardware behavior. As a consequence, low hanging fruit in these areas has already been taken. On the contrary, application-level software design has been mostly unaffected by energy considerations.

Wirth's law [45], formulated in the 1990s, criticizes a reduced attention for efficiency considerations in common programming practice up to a point where dramatic improvements in hardware performance are often canceled out by ever more inefficient software. Whether or not things have been improving concerning time performance, at least it is accepted as an important aspect of software system quality. Not so with energy. Energy had until recently not been considered a scarce resource at all in computing. As a consequence software engineers have not been educated to manage a system's energy consumption and only very limited support is available for the software developer to aid in design and implementation of energy efficient programs. The International Organization for Standardization classification ISO 9216 and its successor ISO 25010 [21] aim for a standardization of all software quality attributes. But not surprisingly they did not even mention energy efficiency as a quality aspect.

This absence of awareness makes it likely that there are big opportunities for optimization. Many authors [29, 9, 31, 11, 6] are convinced that the application software level will get more attention in the next few years because optimizations at this level are believed to have a high impact on overall system energy consumption.

High-level application software has information that is not available at the lower regions of the solution stack. They are closer to the end user and know which tasks are important. Moreover the high-level software architecture determines fundamental aspects of the behavior of the system. In contrast, system optimizations at the data center can be difficult because there is often no knowledge about the processes that are running inside the IT equipment. As a consequence, all computations must be treated equally important and there is little room for trade-offs or changes to the overall system design.

In order to see what is needed to change the situation, consider a software architect who has been entrusted with the task of building an enterprise application that is energy efficient. Since he/she has not been trained in energy efficient design and probably none of his/her teammates have either, he/she will resort to literature. But no books exist that target energy efficient software application design and in general literature on the subject is scarce. Only a few tools exist, mainly not in a production release yet and badly integrated into his/her normal environment. Setting up a test environment and energy monitoring for production will be challenging if not impossible due to the fact that the environments are managed by another organizational unit.

The contributions of this chapter are threefold. First we embed the field of application software energy optimization in a broader taxonomy of sustainable computing or "Green IT," second we present some central problems and concepts, and third we give an overview of work that is available in the literature. In doing so, we have tried to avoid going into technical details or striving for completeness, but rather present a diversity of approaches.

## 2. GREENIT AND APPLICATION ENERGY EFFICIENCY

Since the main purpose of this chapter is to justify the existence of a separate discipline of sustainable computing, we felt it is necessary to give a short overview of various existing activities labeled as *"GreenIT"* and point out how our subject relates to these approaches.

### 2.1 A Taxonomy of GreenIT

Green and sustainable computing or GreenIT has received much attention in recent years. It has become evident that the support of our fast growing

IT infrastructure requires a vast amount of resources and has a considerable negative impact at various levels. The image of computing as a "clean" activity has been confronted with growing concerns about its sustainability. Examples are:

- The often cited EPA report [16] bringing to the widespread attention that IT and especially data centers are major consumers of energy and have impact on national total energy household and greenhouse gas emissions.
- Issues with toxic waste originating from production and disposal of computing equipment.
- Consumption of energy and scarce materials for IT equipment production and transport.
- Limitations of energy provisioning and heat dissipation at large (data centers) and small (processors) scale.
- Electricity costs in operations have become larger than hardware purchasing and management costs in large-scale IT installations.
- The proliferation of wireless and mobile computing increases the importance of reducing the battery usage.

Beside these problematic aspects of direct IT use, there exists a second motivation for sustainable computing. IT can be used to increase the efficiency and the sustainability of other activities. Examples include the research and production of advanced materials or the streamlining of production processes.

A third perspective on its importance comes from the fact that computing has become an integral part of our daily life and has deep effects on our behavior. As such, the design of IT systems has the power to enable or limit a sustainable way of life on the individual or social level. Examples include a software application that stimulates changes in behavior by visualization of personal energy profiles or IT infrastructure that enables teleworking.

As a result we can observe a growing number of very divergent initiatives that have been labeled as green or sustainable computing. They often aim at very different aspects of the relation between computing and sustainability and it can be quite confusing to understand what a specific initiative is directed at and how they relate to each other. In order to reduce this confusion we propose a taxonomy of GreenIT to position the different initiatives and shed some light on their relations (see Fig. 1).

Actually Fig. 1 only displays part of a complete taxonomy, especially those that are relevant in the context of defining the subject of application energy efficiency. Going from right to left in the figure, we are interested in how the *application software* influences the *energy consumption* of an IT system in its *use*

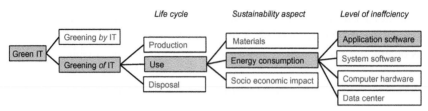

**Fig. 1.** A taxonomy of GreenIT.

*phase.* This is opposed to the question how IT can be used for improvement of the sustainability of other human activities. In the following subsections, we will discuss the perspectives that structure this taxonomy and while doing so, point out what the focus of the current chapter is.

### 2.1.1 Greening-by Versus Greening-of

Initiatives of GreenIT can be distinguished as to whether the purpose is to improve the "greenness" of a computing facility itself (*greening of IT*) or whether it is examined how IT can be used to improve the "greenness" of other processes (*greening by IT*). An example of the former would be the redesign of an operating system such that it allows the server to enter sleep modus more often, to develop a particularly energy efficient memory hardware or a cooling installation for data centers. An example of greening by IT is the use of computation for more efficient Transport logistics or better power grids.

This chapter is concerned with Greening-of IT as its focus is the contribution that software makes to the energy consumption of IT systems. This does not mean that greening-by IT is less interesting to investigate. In fact, since the energy consumed by IT is only a small part of all energy consumed by human activities, applying more IT for the reduction of energy consumption of non-IT processes is estimated [1] to have an even bigger saving potential in absolute terms.

### 2.1.2 System Consumption "in use" and Other Phases of Life-Cycle Analysis

Looking at hardware or software as products gives rise to the question about the sustainability of all phases in the product lifecycle. For the manufacturing and transport of computer hardware a considerable amount of energy and other resources is spent. The same holds for software. It would be interesting to calculate the energy spent on software development and maintenance and compare this to the energy consumed by the system in the use phase. It may very well be that in some cases modern software development environments with their emphasis on continuous build and testing use more energy than the system in production.

It should be noted that the life cycle analysis (LCA) of software is beyond the scope of this chapter. For now, we are only interested in the use-phase energy consumption of systems.

### 2.1.3 Energy Efficiency and Other Aspects of Sustainability

The expression *green and sustainable* is rather vague and can cover a wide range of aspects:
- Toxic waste.
- Labor conditions.
- Electricity consumption (and associated $CO_2$ emissions).
- Other forms of energy consumption.
- Use of scarce materials.
- Socio-economic impact (distribution of wellness).

We are interested in energy efficiency as opposed to one of the other aspects. It is worth noting that many relevant contributions in the search for principles of energy-aware computing are not directed primarily toward sustainability. Consider, for example, research about prolonging battery life in mobile devices, the reduction of electricity costs in data centers, or addressing heat dissipation problems in processors. They develop techniques for energy reduction, but their motivation is not necessarily sustainability.

In this chapter we will focus on electricity use as the dominant form of energy used by IT equipment. Even though accounting and legislation about environmental impact of the industry is often using $CO_2$ emission as central unit of measurement for "greenness." We think that electricity is a more suitable metric for the analysis of IT system sustainability. It is a "normalizing" unit allowing for direct comparison of most of the literature. Relevant articles usually express the system behavior in terms of electrical power consumption and not in terms of greenhouse gas emissions. Some contributions express effects of optimizations in terms of greenhouse gas emissions, but this is typically mediated through the prior calculation of electricity usage. So from the point of view of assessing the "greenness" of systems it is most of the times sufficient to look at their (electric) energy efficiency alone.

A few exceptions exist. One example is the development of routines for workload placements according to the characteristics of electricity generation [47]. But since it seems a good idea to treat energy as a scarce resource even if it is generated in a sustainable way and labeled as *renewable energy* we think it makes sense to strive for energy efficient applications as a goal by itself.

Other aspects like toxic waste, use of scarce materials, labor conditions, and socio-economic effects are also out of scope here. These may be relevant topics for a holistic view on sustainability. However, we feel that these subjects are too broad to serve as a starting point for practical research on software. See Naumann et al. [38] for an attempt to fit these aspects into a project of *sustainability informatics* in the broader sense.

For the remainder of the chapter we will use the terms *energy efficient* and *green and sustainable* interchangeably keeping in mind the explicit restrictions we made here.

### 2.1.4 System Energy Efficiency and Process Improvements

The current chapter focuses on energy efficient systems and not on energy efficient work processes. This implies that we are interested in how to build a system that can deliver a certain amount of messages per energy unit, but not whether these messages are required by the overall business process. In other words, we take the business and functional requirements of the software as given, and try to optimize the energy efficiency of the software within these borders. (In much the same way, optimization of data center energy efficiency takes the application landscape as a given border condition.)

Of course it is very relevant to evaluate the use made of the system and moving to the overall business process level might give insights how to make considerable savings in the IT systems. We prefer to confine the subject here and restrict ourselves to the level of the design of the technical system itself. From the point of view of the design of energy efficient business processes it is important to have a former analysis of what is the consumption of its constituent parts and in this sense it is based on the former analysis of system energy consumption.

### 2.1.5 Efficiency Improvements and Rebound Effects

W.S. Jevons observed in the 19th century that the technical improvements of a more efficient coal driven steam engine triggered an increase of coal usage by the industry [46]. This "Jevons paradox" describes a rebound effect: the fact that increased resource efficiency, when looked at a bigger population, often results in an overall increase in the use of that resource. This may also be applicable to IT energy consumption. The fact that technology advances allow to accomplish a specific task with a smaller energy budget can be seen as one of the main drivers of success of some computing platforms and the resulting increase in energy consumption by this technology as a whole. Consider, for example, my 15-year-old daughter. What will happen when the current functionality of applications on her smartphone could be delivered using only half of the energy? Will energy used by the device diminishes over time? Probably not. Rather applications will either increase features and quality of service until the acceptable energy budget has reached the limit, or she will make even more use of the phone. Both reactions will drive the further proliferation of the platform and increase its aggregated energy consumption. Rebound effects are an interesting line of thought that point out the importance of considering issues at a higher level. But we think that these rebound effects can and should be treated as a separate issue from the technological question of how to achieve a more energy efficient computing technology in the first place.

### 2.1.6 Applications in the Chain of Loss

As mentioned in the introduction, we focus on the application software level because design decisions at this level have deep impact on the efficiency of the system. Even more so, because the area is not so well studied compared to the other system levels like the hardware or the operating system and the device drivers.

The application software level can be imagined as an element of a *chain of loss*. This metaphor looks at a chain of elements that are involved when energy is spent for a computing task. These elements include the electricity generation and its transport; the data center cooling and power provisioning as well as the subcomponents of a server. Each part of this chain has its own inefficiency such that only a part of the input energy is delivered to the next element in the chain. Even the software layers can be seen in this way when we think of them as not consuming energy but computing resources from the underlying hardware or software level. There is overhead and inefficiency involved in every layer of the software stack. Figure 2 gives a conceptual view on such a chain of loss (The proportions are illustrative and do not represent actual measurements). On every level a percentage of the resources is lost and only the remainder is available to the next level. Sustainable computing addresses the efficiency of all these elements of the chain, but the focus in this chapter is on the efficiency at the application level. A well-known metric for the efficiency of the layer labeled "Data center infrastructure" is the *power usage effectiveness* (PUE), but there is no similar accepted metric to express inefficiencies in the software layers.

The energy consumption of the elements in the chain adds up through the whole chain. Viewed reversely the chain of loss appears as an amplification

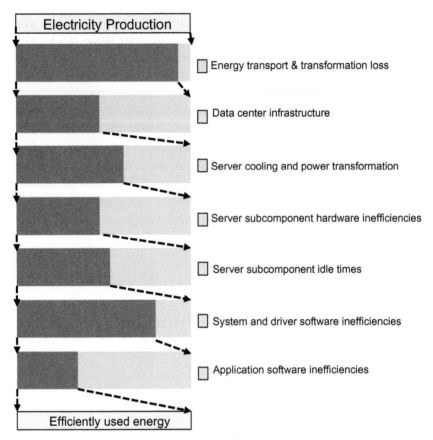

**Fig. 2.** Inefficiency in computing as a "chain of loss."

where small resource consumptions at a higher level (the application soft-ware) will result in larger energy consumption in layers further down the chain. For example, the cooling and power unit of servers or the battery charger for mobile devices add additional consumption to the energy directly consumed by the CPU.

When we focus on absolute numbers rather than percentages of loss it seems often more interesting to address problems where big amounts of energy are consumed (e.g. the data center infrastructure). But in terms of percentages, it may be very rewarding to care about seemingly small improve-ments in the higher layers that are amplified to substantial consumptions through the chain of loss. Figure 3 depicts the same hypothetical system as Fig. 2, but now the losses on each layer are represented in terms of absolute resource consumption instead of loss percentages. Optimizing the software layers seems less attractive in this representation.

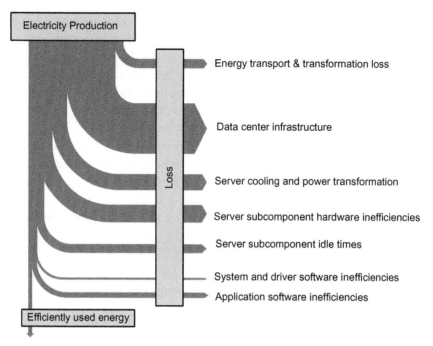

**Fig. 3.** Inefficiency in computing as an energy flow chart.

## 2.1.7 An Organizational Perspective on Applications

There is a slightly different meaning of "application-level optimization" that understands the application more by its functional boundaries rather than as a certain layer of abstraction. An application in this sense is an organizational unit for planning, architectural design, development, and use. The application is a natural entity for many stakeholders to think about energy consumption: The user might want to know the energy consumed for completing a certain task or choose one application over another to provide some desired functionality; the owner of the application might want to know its $CO_2$ footprint or its total cost of ownership (TCO); and for the architect it is a natural scope for design considerations and resource planning.

Optimizations on data center or hardware level only require a very limited insight into the social or organizational structures that determine IT systems. This is often an advantage because it allows applying them directly and at a large scale. For example, in a colocation data center an improvement in cooling infrastructure can be applied to the whole data center without having to bother with details of specific applications and the organizations behind them. The downside is that design considerations and incentives for improvement will be limited. For example, colocation data centers often do not bill application owners proportional to their energy usage and by doing so create no incentive for improving the energy efficiency of the application design. In contrast, large-scale Internet applications that span large parts of data centers or even require dedicated data centers, typically have optimized all layers of their systems. Efforts for improvement are not only directed towards the data center infrastructure but as well towards the high-level software.

Another incentive for change can come from the user. The end user of IT systems cannot choose an efficient data center, but chooses functionality and applications that offer this functionality. When energy usage is reported on the level of applications, the behavior of end users can give additional incentives for improvement. We believe that a focus on the application as a functional and organizational unit can be an important driver for coherent and effective optimization of IT systems.

### 2.1.8 "Software Energy Consumption" and Biased Views on Resources

We want to make a short remark on the idea of software energy consumption as it might occur as a strange concept to the reader. This article is concerned with the energy consumed by computing infrastructure and we are especially interested in the role of software design and implementation for the energy behavior of the total system. Energy, in the form of electricity is of course consumed in the system hardware components, but it is used *on behalf* of the application that delivers some end-user functionality. The amount of energy consumed is influenced for a large part by characteristics of the software. In this chapter we will often refer to the energy *consumed by the software* keeping in mind that software by itself does not consume energy directly.

In order to determine application energy consumption we have to *attribute* hardware energy consumption to software. The mapping of required hardware resource to software is not trivial and depends on the exact motivation for an energy efficient system. There exist different answers to the question: "What are the resources consumed by the application?" If the intent is to reduce the battery operation time for a mobile device, the resources to consider are restricted to the mobile device itself. Even though this might imply changes to the server side application software. In fact, a valid optimization strategy here might be to offload computational tasks to the server and simply ignore the energy consumption on the server side. If the problem is heat dissipation or energy costs in the data center, then the client device energy consumption will not be of interest, but only the servers in the data center and possibly the network equipment and other infrastructure.

Regarding sustainability or "greenness," it makes sense to look at *all* resources consumed by an application. That would include, for example, the client and the server, and the networking infrastructure in between. We think that even while often such a holistic view is not necessary and probably not even practical in many situations, it is illustrative to read the existing literature from this angle: what part of the system in question is in the scope of the inquiry and which parts are omitted. Energy optimization strategies are based on these choices for which resources can be omitted and the result might be called energy efficient, but not necessarily "green." Sometimes it is useful to break the problem down into analysis of parts. However, it is important to keep in mind that from a sustainability point of view these parts should be put together at a certain point to judge the effectiveness of the approach. Application level energy efficiency optimization as a discipline of Green IT wants to know how efficient certain functionality for the end user is delivered by the system and it does so with a rather global view on the underlying resources.

# 3. APPLICATION-LEVEL OPTIMIZATION AS A DISTINCT DISCIPLINE

There is already a considerable amount of literature that is directly relevant for optimization of application energy efficiency. The main focus is often not the application software level, but more generally the software level, meaning the system software or operating system level in many cases. Nevertheless, we think that inquiry into application-level energy efficiency is currently evolving as a separate discipline of the science of power management. The field of research is at a stage that an overview of contributions makes sense given the increasing number of articles published and the appearance of dedicated workshops on the subject matter.

In Table 1 we present contributions from the literature that we think are especially insightful or representative for the topic. We do so by dividing them into three groups that represent the type of support they give to the reader in the analysis and optimization of energy efficiency of software applications:

1. *Conceptual:* These texts focus mainly on survey of existing work and clarification of central concepts and principles. These concepts are either reoccurring in other literature or are novel and describe fundamental aspects of the problem.
2. *Tooling:* These are texts that focus on supporting measurement and modeling of systems under the aspect of energy efficiency. Hence tooling is

**Table 1** Literature categorization.

| Title | Reference | year | Conceptual | Tooling | Practice |
|---|---|---|---|---|---|
| Towards Understanding Algorithmic Factors Affecting Energy Consumption | 22 | 2005 | ■ | | |
| Energy-driven statistical sampling: detecting software hotspots | 12 | 2002 | | ■ | |
| The Case for Energy-Proportional Computing | 4 | 2007 | ■ | | |
| Fine-Grained Energy Profiling for Power-Aware Application Design | 24 | 2008 | | ■ | ■ |
| Delivering Energy Proportionality with Non Energy-Proportional Systems - Optimizing the ensemble | 35 | 2008 | | | ■ |
| A comparison of High-Level Full-System Power Models | 31 | 2008 | | | |
| Energy efficiency: The new holy grail of data management systems research | 19 | 2009 | | | ■ |
| Towards a science of power management | 25 | 2009 | ■ | | |
| Energy Consumption in Mobile Phones: A Measurement Sudy and Implications for Network Applications | 3 | 2009 | | | ■ |
| pTop: A Process level Power Profiling Tool | 14 | 2009 | | | |
| A Framework for Estimating the Energy Consumption Induced by a Distributed System's Architectural Style | 33 | 2009 | | | |
| Analyzing the Energy Efficiency of a Database Server | 34 | 2010 | | | |
| Recipe for Efficiency: Principles of Power-Aware Computing | 30 | 2010 | ■ | | |
| Energy-efficient algorithms | 2 | 2010 | ■ | | |
| Power-Efficient Software | 32 | 2010 | ■ | | |
| Toward energy-efficient computing | 9 | 2010 | ■ | | |
| Towards a Model of Energy Complexity for Algorithms | 22 | 2010 | | ■ | |
| Power Measuring and Profiling: State-of-the-Art | 13 | 2010 | | ■ | |
| Virtual Machine Power Metering and Provisioning | 23 | 2010 | | | |
| Greenalytics: a tool for Mash up life cycle assessment of websites | 27 | 2010 | | ■ | ■ |
| SPAN: A software power analyzer for multicore computer systems | 37 | 2010 | | | ■ |
| Architecture and Mechanisms of Energy Auto-Tuning | 18 | 2012 | | | |
| Is software "green"? Application development environments and energy efficiency in open source applications | 11 | 2012 | | | ■ |
| Who killed My Battery: Analyzing Mobile Browser Energy consumption | 54 | 2012 | | ■ | ■ |

taken in a broad sense of offering support in the task of designing and evaluating Software systems. A common goal of these texts is to grasp the energy behavior of (application-) software as a basis for reasoning about efficient design, optimizations, and for further research.

3. *Practice:* In this category falls all texts that mainly present a particular optimization approach or that present concrete results for energy profiles related to some aspect of a system or a class of systems. In this class falls also contributions that formulate general recommendations for practitioners like software architects and system developers.

It should be noted that many texts make contributions to more than one aspect, especially since the field is quite young. For example, attempts to develop a certain profiling tool will often be based on a slightly new conceptualization of the problem. This might be explicit, by the presentation of a new designed metric, or in a more implicit manner. Also many contributions in the "tooling" section present some findings on concrete optimizations to illustrate their approach and demonstrate its correctness.

The next three sections are organized along these lines: first we will present some basic concepts and principles. Then attempts for metrics and tooling are presented, followed by a section on optimization approaches.

## 4. CENTRAL CONCEPTS AND PRINCIPLES

Before we go on with presenting some approaches for measuring, modeling, and optimizing application energy efficiency, we would like to explain some basic concepts of the field that will help in understanding these approaches.

### 4.1 Energy is Not Power

This is quite a trivial physical fact, but in the literature they are sometimes used interchangeably, it is useful to keep in mind when the distinction matters. Electrical energy is power over time. The measurement unit for energy is Joule (J). Electric power is measured in Watts (W), the energy used by electric devices is often simply expressed in kilowatt-hours (kW h). Energy is not measured directly, but power is determined for a certain moment in time. If the power is constant for the duration of the measurement, the energy can be simply calculated by:

$$\text{Energy} = \text{Power} * \text{Time}.$$

For a high-level estimate of a system's energy consumption over a certain period, it is sufficient to know its average power for some representative workload. Alternatively the power from the system's hardware specification is often used. However, when we are concerned with optimizing a system and want to get a more detailed picture, we have to deal with the fact that the power consumption of a system or its subcomponents is not constant over time. Fluctuations can, for example, correspond with changes between running and resting states. If the power consumption is not constant, the energy is calculated by the time integral of power. Apart from requiring another mathematical formula, it reveals the dynamic aspect of energy consumption. From an optimization point of view it is possible that optimization for energy involves increase in power for a certain time; like in the *race to idle* strategy that exploits the better energy efficiency of high activity and idle system states while avoiding intermediate utilization modes with lower efficiency.

The distinction between power and energy can also be reflected in the difference between a *power model* and an *energy model*. The former is a characterization of the system component's power consumption in a certain state or configuration. It is typically used in combination with an energy optimization algorithm by some hardware resource manager to calculate the optimal strategy of execution for a given workload. At a higher software level, optimization strategies may not require power models, but energy models do because important knobs to manipulate low-level behavior of the power consuming device may be encapsulated by a component or service interface. Decisions on detailed state switches (and related power levels) are taken autonomously by the component. A consumer, that may be a human or another process, is then guided by total energy used for a given computing task, not by the power states of a device. An example of this is the labeling of services with energy consumption metadata in a service-oriented environment or the high-level workload placement as proposed in Götz et al. [18].

## 4.2 Consumption is Not Efficiency, Activity is Not Productivity

Assume that we know the energy consumed by a software system in some time period. This by itself can be valuable information in some cases. For example, when we want to know which part of a system composed of multiple components are most interesting candidates for further optimizations. But generally energy efficiency requires some measure of work that is performed using this energy. While for hardware energy efficiency it is sufficient to express this measure in some well-known low-level unit, e.g. instructions per second for a CPU or bytes per second for a network device. This does not hold for the application software layer. Its efficiency cannot be expressed in this way, because often the strategy of energy efficiency improvements at this level is lowering the total energy spent by reducing the amount of requested resources. These efficiency gains will not be observable when looking only at low-level resource activity, because the aim is to reduce the consumed resource-level units while delivering the same high-level work unit. Or putting it differently: an active system does not mean that something useful is delivered to the user

$$\text{Energy Efficiency} = \text{Useful Work}/\text{Energy}.$$

In order to express application software energy efficiency and total system energy efficiency from the perspective of the end user, the performed work must be expressed as a *functional unit of work* that is delivered at the system boundary. This can be something like number of transactions processed, user requests responded to, amount of multimedia content delivered, or contracts entered through the system. A problem here is that functional units of work that are interesting from a user's point of view, like playing a dvd, making a phone call, sending an email, or making a search by a search engine, are typically only meaningful for a very restricted class of applications. This limits the way that results can be compared between applications. See discussion of Carpa et al. [11] and Intel's EnergyChecker [15] that are shortly discussed in the next section for attempts to deal with this difficulty.

## 4.3 Idle: It's Getting Interesting When Nothing Happens

The concept of resting or being *idle* plays a central role in the literature of energy efficient systems in two ways. First for optimization: since it is very common for a certain component of a system to stay in an idle state, it is important that the energy consumed in that state is reduced to a minimum because it does not do anything useful by definition. Second, for the measuring of energy consumption it is important to make a distinction between idle and active states since it supports the analysis of the software and hardware behavior. For example, some process is apparently not consuming a resource if it is in an idle state and in this way the detection of the idle state allows to better observe software induced resource usage. Model-based estimation approaches typically make the distinction between idle and active states by attributing them different power levels. Optimization algorithms then use these power models, together with information on *activation costs* to choose the best way of execution.

### 4.3.1 Idle Consumption: System Baseline Versus Application Idle

Because components, system, and whole computing infrastructures use energy when in idle state, better results for measurement and estimation can be obtained when the idle consumption is known. With respect to software energy consumption it is interesting to make a further distinction concerning the system idle consumption:

- *System baseline consumption:* The consumption of the system without running the application.
- *Application idle consumption:* The consumption of the system while the application is running, but not doing anything useful.

Just imagine you start up your desktop without any programs running. This is the baseline consumption. Now open your preferred text editor but do not type anything. This is the application idle consumption. When you interact with the application or the application does some background processing this is active consumption. It is not universally defined what processes are counted as baseline consumption or which activity of an application counts as performing useful work, but that normally gets clear from the specific context.

### 4.3.2 Multiple Device Power Levels

A model that describes a resource as either used (active) or not used (idle) is a two-state power model. Of course things are not that simple. In order to reduce energy consumption two mechanisms are applied widely: allowing the resource to enter different sleep states that allow it to consume even less power (but incur probably a wake-up cost) and trading performance for energy by allowing the resource to be active but work at a lower performance and power level. The states are called *power states* and *performance states*, respectively, and have been formalized, for example, in the Advanced Configuration and Power Interface (ACPI) standard [37]. An example of the use of performance states is the dynamic frequency and voltage scaling (DFVS) approach for the processor.

The resource states can be used by a resource manager to minimize the energy consumption while keeping an acceptable level of performance. Most of the literature that is concerned with device states and resource management is about low-level implementation in the operating system and driver level, but high-level application programming constructs like continuous polling loops can have a big impact on device behavior and a basic understanding of device states is needed for analyzing the overall program behavior and the detection of energy bottlenecks.

## 4.4 Energy Proportionality and Utilization

Barroso and Hölzle have coined the term *"energy proportionality"* in a seminal paper in [4]. Their key contribution was that besides minimal power consumption at a given utilization of a component another property is fundamental for the design of energy efficient systems. The component should be able to scale its energy consumption according to the utilization such that when not used it will also not consume energy *and* the energy consumption should grow proportional with the utilization. The importance of this property stems from the fact that: (a) low utilization is not the exception, but rather the normal operating mode for many platforms, (b) low utilization is often by design (as results of trade-offs for reliability and operational maintenance), and (c) energy-proportionality (also called *energy scaling*) of today's hardware components, systems, and large-scale computing infrastructures is very limited. These three factors make that current computing systems are far less efficient simply due to the fact that they do not behave energy proportionally.

An important observation in the context of this chapter is that the concept of energy proportionality can also be applied to software applications—i.e. a system should reduce its energy consumption according to the amount of work it is doing. The only thing to keep in mind is that hardware utilization is not a good measure for the amount of work done by a software application. As we stated above, activity is not efficiency and hence at the level of application software the consumed energy must be related to a unit of useful work from the user perspective.

## 4.5 Performance and Energy Efficiency: A Many-Faceted Relationship

Energy efficiency and performance are closely related to each other. This can be seen from looking at their equations:

$$\text{Performance} = \text{Work}/\text{Time},$$
$$\text{Energy Efficiency} = \text{Work}/\text{Energy}.$$

Both properties express the efficiency with respect to a certain resource: time or energy. This makes it clear that at least in principle those interests might be in conflict. It is not clear though in which situation this potential conflict does manifest itself in practice and is relevant for the process of system optimization.

### 4.5.1 Two Basic Optimization Strategies

Since in many contexts there are restrictions on both time and energy, it is interesting to know if and when the optimization for one property conflicts with the optimization of the other.

On the processor circuit level, performance and power have a quadratic or even a cubic relationship (which is exploited in DFVS). And many energy saving schemes rely on the principle of trading energy intensity for performance. On the other hand, the observation of bad energy proportionality of current server subcomponents and the whole resulting systems [4] suggests that the observed trade-off between time behavior and energy usage does not necessarily translate to a trade-off at the system level. This holds at least for the majority of currently deployed server platforms. Today's servers' energy efficiency is best at high-performance states. Or to put it differently; as long as the energy consumption does not scale with the work done by a system, improving the energy efficiency can best be achieved by increasing the performance of a given system. Similarly, Tsirogiannis et al. in their analysis of a database server energy behavior [35], point out that optimization of systems for performance will deliver energy efficient systems automatically. This suggests that the difference between performance optimization and energy efficiency optimization is unimportant in situations where constant high utilization of resources can be achieved.

We extract from this discussion two basic strategies for improvement of energy efficiency of a system:

1. *Avoid low utilization:* Maximize utilization and throughput, and minimize energy consumption at high utilization.
2. *Aim at energy proportionality:* Embrace low utilization and minimize energy consumption at all levels, especially at idle/low utilization.

An energy efficient system can be optimized using both strategies, but this is not a necessity. If the system design can achieve high utilization at all times, for example, by workload scheduling and fine-tuning for the high-utilization mode, there is no need to additionally care about energy scaling.

An interesting border case is the ensemble level optimization where energy efficiency strategies cover not a single device, but multiple nodes that do similar work. Here, overall system proportionality can be achieved by switching off components that have by itself a non-energy-proportional behavior [34].

Another border case is the optimization of energy consumption for the most common case. Here, neither a high utilization is targeted, nor does the system design aim at general energy proportionality over all utilization levels. Rather the statistically dominant usage level is identified and energy consumption is optimized for that, accepting less efficiency at low and high utilization levels. If the amount of time spent in the optimized range is large enough this strategy can be a valid approach to optimize overall system efficiency.

# 5. TOOLING: GETTING GRIP ON APPLICATIONS ENERGY BEHAVIOR

You can only manage what you measure. Characterizing and improving the energy efficiency of application software requires first getting some data about energy consumed by the software. As we already mentioned, energy is not consumed by software, but by hardware on behalf of software. So there is always an indirection involved. You cannot put your meter directly into the application. The simplest approach is to connect the meter to the hardware running the software with some specific workload and say that the energy used is taken as the energy consumed by the software.

This basic strategy may lead to usable results in some situations. For example, if you are interested in the consumption of a single embedded device that runs exclusively some dedicated software. It may be useful also for a comparative study as reported in [8]. But for many instances concerning software-related energy consumption on current computing infrastructures this setup is too simplistic.

In the next subsection, Section 5.1, we will discuss about the needs of advanced techniques for determining software energy consumption. Modeling and measuring approaches for application energy behavior should address at least some of the issues mentioned here. Section 5.2 presents some of the approaches in the light of the "requirements" formulated in Section 5.1.

## 5.1 Motivations for Some Desired Properties of Models and Metrics

### 5.1.1 Spatial Distribution of Resources

A software system can be distributed across multiple hardware nodes. In fact, in today's networked environments, this is the standard case rather than the exception. Just consider virtualization, clustering architectures, and collaboration of application software components deployed on different devices and storage networks. Which nodes of the system landscape should we take into account in order to determine the energy consumption of the software in question? Where software uses a larger amount of spatially separated resources, the question of system boundaries and scope of interest becomes relevant. An energy model for application software might require the collection and correlation of data from very different locations and sources.

### 5.1.2 Attribution of Shared Resources

Not many platforms run only a single application at the same time. This holds not only for today's common mobile and desktop platforms but also for many server configurations. On a desktop you can, for example, write a document while you listen to some music. In the same way, a server often handles different tasks simultaneously. This sharing happens not only at the level of the processor, but also in the other system subcomponents like memory, disk, and network interface.

Sharing of resources has a twofold effect for the determination of software energy consumption: First, in order to determine the energy used by a particular application, the total system power measured at the hardware device level must be *attributed* to that application by separating it from energy consumed on behalf of other applications. Second, the behavior with respect to resource usage of some piece of software might be influenced by activities of other programs running. This in turn can have an effect on the energy that it consumes. So while it will normally be of interest to determine the energy usage in a typical situation, this typical situation is often characterized by sharing of resources and requires dividing up total system energy consumption. Hence tools that support the assessment of software energy efficiency should be able to attribute resource usage to different processes in a shared environment.

### 5.1.3 Interdependencies with Other System Layers

We are interested in application software energy consumption and a natural way to go is to look at the effect that modifications to the application design, implementation, or configuration have to the total system energy consumption *all other things being equal.* But of course other things are not always equal and we would also like to have results that are portable and valid on a changed environment. That means changes at the hardware sub components or their configurations, the operating system, or the workload. A mobile app should ideally be able to run in an efficient way on many different mobile devices. A server application might be moved to a virtualized environment. A change in storage technology or in operating system settings can have deep impact on the system energy efficiency and undermine the portability of energy measurement results and models. The energy consumed is dependent on all those factors and the specific combination of system parts at all different levels. In a similar way changes in the workload, like an increase in users or the amount of data it has to manage, can change the energy behavior of a system.

A software energy model should be accurate, yet allow for a certain amount of portability to other platform configurations. For energy efficiency optimizations it is relevant to know whether they are generic or based on specific platform properties.

### 5.1.4 Cross-Application Comparability

A metric for energy efficiency does typically allows *self-comparison* by re-measuring the same system with different variations. Variations can be, for example, changes to configurations, workloads or the design, and implementation.

A user faced with multiple alternative applications offering certain functionality might be interested in knowing which of the alternatives can fulfill his/her needs with the lowest energy budget. An application owner or designer might want to know how the energy efficiency of his/her system compares to other similar systems. These are natural questions that are not unique to the energy efficiency aspect of systems and a whole industry is devoted to benchmarking and comparing computer systems with respect to all kinds of properties.

Ideally a metric for application energy efficiency should offer a way to compare the results for different applications, at least within a certain class.

## 5.2 Overview of Approaches for Modeling and Measuring

This section addresses some approaches for energy measurement and modeling that aim at supporting application-level optimizations.

### 5.2.1 Energy Complexity of Algorithms

Since energy can be seen as just another resource of computation similar to, for example, time, attempts have been made to model energy behavior of algorithms along the lines of time complexity and the big-O notation that has been proven to be very useful for performance analysis of algorithms. The *energy complexity* of an algorithm is a way to characterize its energy needs in a fundamental way, independent of a specific platform or implementation. If parameterized with platform characteristics, this would allow to define upper or lower bounds on the energy that should be needed for a certain computation. This could support choices at the algorithmic level during the design time or maybe even allow comparing implementations to a theoretical optimum. The underlying assumptions is that the energy complexity at the algorithm level has a significant influence on the total energy consumption of real applications or even that the energy behavior of real applications can be modeled bottom–up from the energy complexity of the algorithms of the application. Martin [39] is an early example of an attempt to formulate an extended concept of algorithmic complexity covering time as well as energy aspects. His discussion is however concerned with algorithmic choices for the integrated circuit. Jain et al. [22] attempt to model the energy complexity of algorithms at the application level by their "switching complexity," but, besides the theoretical considerations, their result is largely negative, as they do not succeed in validating the model with experimental evidence. An important point from considerations on energy-complexity of algorithms in the literature is that it makes no sense to treat energy consumption in isolation from time considerations. The energy consumption of a sorting algorithm does not only depend on characteristics of the input (data size and randomness), but also on constraints on the performance of the algorithm. An algorithm may be more energy efficient if it is allowed to take more time to do its work.

### 5.2.2 Hardware-Based Power Profiling

We presented the most simplistic setup for device-level measurements using an external power meter at the beginning of Section 5. There are of course more advanced setups that measure power at hardware subcomponent level like disks, memory, network, or processor, or even inside a subcomponent. Aside from requiring system-specific hardware instrumentations, it is a non-trivial task to attribute the consumed power to specific applications or even application subroutines. The difficulty of instrumentation is partly resolved by the introduction of integrated power sensors into newer hardware. The problem of attribution to software activity can be addressed by correlating system activity and power metrics by taking fine-grained samples of instantaneous power consumption together with process information. The derived energy during a sample interval is then attributed to the active process at the sample time. An interesting discussion of such statistical sampling can be found in Chang et al. [12]. They propose not to use time as the trigger for sampling, but the energy consumed by the system itself. Every time a certain amount of energy has been used by the system, a sample from the process activity is taken, which results in samples of varying lengths. An advantage of this approach compared to straightforward timer-based sampling is that the sampling rate in low activity and idle states will be slower and measurements will be less interfering with the system to be measured. They demonstrate on a mobile platform that this indeed leads to a higher accuracy. Ge et al. [17] propose a more detailed way to establish a semantic relation between power dissipation and program structure and demonstrate the method in a high performance distributed multicore environment. In addition to the direct power metering at system component level, they use simultaneous logging from application code instrumentation to better associate power consumption with program flow.

Hardware measurements are very accurate but it remains difficult to correlate metrics to higher level software structure, especially in shared and distributed environments. If no integrated metering is available they require physical access to and often modification of the machine, which severely restricts their applicability in many real-life production situations. Hardware measurements are however very important for the validation and calibration of software-based methods that are discussed in the next subsection.

### 5.2.3 Indirect Measurement Using Resource Counters

The key idea of this approach is not to measure the power consumption of the hardware components directly, but instead read out readily available system *performance monitoring counters* (PMCs) and calculate the current power consumption using a *power model* of the system hardware. One advantage of this utilization-based approach is that the power model can be determined offline and once determined, supports run-time estimation of power dissipation. These run-time measurements can be completely *software* based and do not require a hardware instrumentation. It suffices to monitor the readily available counters. Another important advantage is that this allows in principle to calculate the energy used per-application, since the performance counters can usually be attributed to specific processes. Bellosa [5] is an early example of using this method to determine power usage of the CPU on a per-thread level. Rivoire et al. [30] have compared PMC-based approaches with other simple ways to estimate power at the full-system level observing that counter-based energy metrics yield highly accurate results even with a very limited set of counters that are commonly available in many platforms. Being able to use a restricted set of ready available counters increases the wide applicability of this technique. However, in the light of fast evolving hardware component characteristics, the accuracy of a specific set of proxies for whole-system energy consumption, as reported in [30], cannot generally be assumed and should be reestablished for the platform in question. Kansal and Zhao [24] and Do et al. [14] extend the concept of PMC-based energy consumption estimation to the development of real-time process-level profiling suited for optimization of the application-level design. The research culminated in the profiler applications Joulemeter [40] and pTop

[41], respectively. Both profilers target the Windows platform, but the principle would be applicable to other operating systems as well. Kansal et al. apply software-based power metering to virtualized environments in [23]. Virtualized machines are a difficult platform for power metering because hardware instrumentation on the single machine level is not possible and sharing of hardware resources by different applications occurs by design. Given the fast growth of cloud computing the applicability of PMC-based approaches to virtualized environments is an interesting property.

Performance counter-based measurement is applicable for many platforms. Suitable PMCs that can serve as a restricted set of proxies for total system energy consumption are generally accessible in modern operating systems and do not anymore require system software modifications. A downside is that the approach requires a system-specific calibration that needs to be repeated on any changes in configuration in order to guarantee the desired level of accuracy. This will involve direct hardware power metering.

Another extension to this approach can be found in Wang et al. [36] where power consumption is not only attributed to the process level, but to the function level in source code. However, this useful feature comes at the cost of required source code instrumentation that makes it less applicable in a lot of cases. The article [36] (and a later article by the same author [13]) also contains a good overview on related power measurement and profiling techniques.

### 5.2.4 Relating Energy to Useful Work

We already mentioned that relating the energy consumed to a unit of useful work is necessary to meaningfully assess the energy efficiency of applications. Intel has developed the EnergyChecker SDK [15] that enables monitoring of productivity in terms of useful work units and correlation of that information to energy metrics in a uniform way across applications. It works by application software instrumentation that traces predefined functional events or "counters" and subsequent correlation of those counters with independently measured power consumption data. The latter can be provided either by an external hardware power meter or software-based solution and the correlation and visualization of the data can occur offline or online on a running system. This is potentially a very powerful approach, since it answers to the operational needs of uniform management of a larger collection of applications and enables fast analysis and visualization. A drawback is its limitation to the Intel platform, since at least the current implementation of the approach relies on specific hardware features. Furthermore it requires software instrumentation, which makes it impractical for the assessment of energy efficiency of many existing systems, since in industrial systems code instrumentation is often considered to introduce too much instability. Finally, since the unit of useful work is defined on a per-application base, this approach in principle allows for standardization of these functional counters and as a consequence enables the direct comparison of energy efficiency among functionally similar applications.

### 5.2.5 Measuring the Energy Cost of Abstraction Layers and Encapsulation

Capra et al. [11] make an attempt to empirically confirm the intuition that the use of frameworks and libraries that allow easier development and maintenance should have a negative impact on energy efficiency. The reasoning is that first, frameworks and libraries tend to offer general solutions and can be only optimized in a limited way to the specific application (This point is also mentioned by Ranganathan [29]). Second, the use of abstraction layers causes inefficient execution because excessive layering adds to the execution paths. In order to grasp this intuition, Capra et al. [11] design a metric that measures how much code from external libraries is used in a program. The metric basically counts the number and frequency of features used from a language or application programming interface (API) as opposed to a programming style that confines itself to a limited set of basic language constructs. The authors call this the "Framework entropy" in allegory to Shannon's entropy. Another interesting aspect in [11] is how the definition of a suitable unit of work is addressed. The energy efficiency of a program is defined as how it performs relatively to other programs inside a class of functionally similar applications given a predefined workload. The resulting metric in this way expresses the energy efficiency of a system in terms of how it relates to other functionally similar systems. Capra's model is an illustrative example of how a certain aspect of interest that is suspected to influence the energy efficiency of a system can be made quantifiable and operational.

### 5.2.6 Predicting Effects of the High-Level Application Architecture

Seo et al. [32] developed a model to predict the energy usage of an application based on the "architectural style" of the application software architecture. As example they use some styles of distributed system architectures: *publish-subscribe*, *peer-to-peer*, and *client-server*. The model is designed to be used in early stages of the system development; hence no application is yet available for actual measurements. Instead the approach is purely based on simulation.

Nevertheless the authors show the estimation has a high accuracy compared to the measured energy consumption of the implemented solution.

The estimation works by developing first an abstract *"energy cost model"* for a specific architectural style that describes the nature of communications and data transformations in this style. In a second step, this style-specific model is parameterized with the actual layout of the application in questions (e.g. the number of logical connections between components in the application under evaluation), and the platform-specific parameters (e.g. the measured energy cost of executing one of the activities of the style-specific model on the target platform). The result is called an *"energy prediction model."* When such models for multiple styles have been constructed, the styles can be compared using specific application usage scenarios in order to support the choice of the best alternative.

An interesting aspect of this estimation approach is that it focuses explicitly on high-level application architecture attributing energy footprints at the software component level. Furthermore, it can be used early in the design phase of an application when executable code is not yet available.

### 5.2.7 Observing Website Efficiency by Low Precision Estimation Models

Greenalytics [42] (and the similar, but commercial service Green Certified Site [43]) are services that estimate the amount of $CO_2$ emission caused by websites. Lamela et al. [27] explains the underlying model for Greenalytics which is based on calculating energy consumption from statistics of the website in the Google Analytics service [44], some reference parameters taken from the literature about energy dissipation of internet traffic, and average client device energy consumptions. The results are related to a functional unit of work for the end user, like a website visit. While the model is presumably not very accurate and does take into account the server back-end only in a trivial way, it is an attempt to achieve a high-level energy estimation model that aims at complete coverage of all energy dissipation caused by an application. Different from many other models, it is not solely focused on the server or client site or on a single node. Furthermore, the approach is independent from specific platform details, requires little instrumentation (usage of Google Analytics), and is readily available as an online service. Despite the low accuracy, such an approach can give valuable insights on the energy footprint of a website for stakeholders and may be sufficient to guide user decisions and take the low hanging fruit for optimization of the energy efficiency. We think that similar high-level models and measurement infrastructures are essential for reasoning about the impact of IT from

Table 2 Modeling and measuring approaches.

| Desired Property | Deals with Distributed or Networked Applications | Attribution of Shared Resources to Individual Applications | Broad Portability to Other Platform and Workloads | Allowing Cross-Application Comparison | Accuracy |
|---|---|---|---|---|---|
| Martin [39] | | + | + | | Medium |
| Jain et al. [22] | | + | + | | Medium |
| Chang et al. [12] | | + | | | High |
| Ge et al. [17] | | + | | | High |
| Bellosa [5] | | + | | | High |
| Rivorie et al. [30] | | | | | High |
| Kansal and Zhao [24] | | + | | | High |
| Do et al. [14] | | + | | | High |
| Kansal et al. [23] | | + | | | High |
| Wang et al. [23] | | + | | | High |
| Intel EnergyChecker [15] | | | + | + | Medium |
| Capra et al. [11] | | | + | + | Low |
| Seo et al. [32] | + | | + | + | Low |
| Greenalytics [42, 27] | + | | + | + | Low |

a sustainability point of view. The examples given here must be seen as first attempts and are still in need for refinement, but the basic idea, to trade accuracy for easy applicability, comparability, and a complete view (in terms of covering all resources used by a single application) is very valuable.

The Table 2 lists the aforementioned approaches for modeling and measuring and shows which of the desired properties that have been formulated are present. We also added an indication of the achieved accuracy because that points out some of the trade-offs involved.

# 6. OPTIMIZING

We have seen some examples of how the energy behavior at the application level can be determined and what type of models and measurement setups are proposed in the literature. But how do optimizations for energy efficiency at the application software level work in practice? What are typical inefficiencies discovered? Can sources of inefficiencies and optimization techniques be categorized into groups? How are they detected? What are candidates for "principles of energy efficient system design?" As in the previous section, we will present some promising work in this area. Before we describe some specific approaches for optimization we would like to present two perspectives on the problem that help in categorizing the different strategies that can be found in the literature; first, the distinction between resource consumer and producer efficiency and second, the detection and formulation of inefficiency patterns.

## 6.1 Distinguishing Responsibilities of Resource Consumers and Resource Managers

In a short and interesting article Saxe [31] divides energy efficiency optimization approaches along two roles of software in computing systems: *resource consumers* and *resource managers* (or *resource producers*). Resource consumers are focused on delivering the functionality of the system and have the responsibility to minimize their need for computing resources. Examples of resource consumer optimization would be the reduction of memory allocation by inefficient data structures and algorithms or avoidance of polling loops.

Since there are more and less energy efficient ways to consume a certain resource, specialized software (i.e. resource manager) can determine how the requested resources are used in the most effective way. These resource managers have become an essential part of systems since modern computing infrastructures and the constituent components have a very complex power behavior. Resources are typically only partially utilized and each resource offers multiple alternatives for being used. An example for a resource manager is an operating system routine that exploits different possible power states of the device it manages. Resource management is not restricted to the lower system level and can also be implemented at the application level. An example here would be a peer-to-peer network application that implements energy-aware node selection as described in Blackburn and Christensen [7].

For resource management Saxe identifies at least two possible strategies; "spatial considerations" are concerned with routing the resource request to the best fitting resource or resource state while "temporal considerations" try to group resource requests in time to better exploit resource power states (also called task *batching*). Resource management requires some knowledge about energy profiles of the resources and possibly other context information like workload characteristics and resource state to make a well-informed decision. It also needs information on performance characteristics of the alternatives in order to make correct trade-offs. For example, for efficient task batching, the manager needs information on the acceptable delay of execution. A way to provide the latter is to extend program function calls (resource requests) with information on the required level of service like the maximal delay for some scheduled event. Albers [2] gives a survey on research on energy optimization algorithms that are implemented by resource managers. She also presents some specific results like, for example, the optimization problem of constructing a state transition schedule for a system with multiple power states.

A point worth noting on the formulation of these energy optimization problems is that they always have an implicit performance constraint; it makes no sense to formulate an energy efficiency optimization problem without constraints on the acceptable level of service.

## 6.2 Patterns of Inefficiency

In [29] Ranganathan identifies common reasons for energy inefficiencies. He also mentions common strategies for optimizations. Both are formulated independent of the system level and can occur in the hardware implementation as well as on the higher software level or the data center infrastructure. As far as we know, this is the only attempt to compile such a list of sources of inefficiency that tries to abstract away from the specific area of application in the solution stack. Here is a short summary:

- *General-purpose solutions:* Devices or components often fulfill many functions. This opposes alignment of resource consumption to a specific use case or required quality of service (QoS). The solution will assume the worst case of all possible uses.

- *Planning for peaks and growth:* Design solutions focus on satisfying the high capacity requirements from peaks or expected growth and are not able to scale down resources on low workload. But in practice low utilization is the common case. This leads to poor efficiency in the most common utilization mode.
- *Design process structure:* Often system design is organized by dividing the system into components, layers, or other abstractions. Different engineering teams are focused on different parts of the system. They tend to implement energy optimization with a local perspective. From a global perspective it would often be more efficient to have a tighter integration or alignment between components or layers. This could, for example, be achieved by a better communication between those parts.
- *Tethered-system hangover:* The absence of energy efficiency as a design constraint, the practically unlimited availability of energy in design and production environments, as well as the exclusion of energy costs from the total cost of ownership (TCO) results in energy inefficient systems, especially when other design constraints like performance push in another direction.

Ranganathan [29] continues with providing 10 "categories" for efficiency improvements. They make an interesting alternative starting point for a classification of optimization strategies to the one we give below. Even though there is considerable overlap with the main strategies we present here, we found that the approaches for optimization we wanted to refer to where not fitting very naturally onto them.

## 6.3 Some Optimization Strategies

### 6.3.1 Just Plain Performance Optimization

As already mentioned earlier, performance and energy efficiency do not necessarily require distinct optimizations. For a certain type of optimizations and a certain type of systems and workloads performance optimization will also improve energy efficiency. This is especially true for high-utilized server systems that are optimized for throughput. For example, Harizopoulos et al. [19] have first suggested shifting focus from performance centric to energy efficiency-oriented research for data management systems, but in a later text [35] they conclude that at least for an important class of system architecture (databases on a single, independent node designed for scaling out) "the most energy efficient configuration is typically the highest performing one."

There are two basic intuitions behind this strategy. First, in the face of systems or system components that have bad energy proportionality and hence low energy efficiency in low load regions, augmenting the throughput of a system will increase its energy efficiency. Second, if performance improvements are achieved by reducing the amount of resources needed, for example, by choosing an algorithm that needs fewer steps, it will result in lower energy consumption. But not all situations lend themselves to this approach. For example, smartphones have a fixed energy budget and only limited possibilities to augment the underutilization of resources. Pure optimizations for throughput will probably not be the best strategy to follow here. Performance centric optimization can also have negative impact on energy efficiency if the effort is directed solely toward time behavior improvements and ignores resource efficiency.

### 6.3.2 Don't Stand in the Way of Power Management

Optimization can also take a rather passive attitude towards energy saving. The idea is then not to focus on avoiding bad behavior that undermines energy saving features implemented at other levels. One of the most popular recommendations from the literature (see e.g. [31]) is avoiding polling loops with a high-resolution timer. The reason for this particular case is that it potentially wakes up part of the system from sleep states, while in most of the cases there is nothing to do. This is a typical case where application behavior can undermine lower level power saving routines. The "tickless kernel" from the lesswatts.org initiative [48] follows this approach and Intel's PowerTOP [49] is a tool that helps in detection of bad behaving programs. Another example of disturbance of power saving features is the retention of unnecessary references to data in memory thereby opposing automatic garbage collection and possibly transition of memory to a lower power state.

### 6.3.3 Communicate the Required Service Level

Related to the previous, there is a slightly different strategy that also aims at enabling existing energy saving features. But instead of being focused at avoiding disturbance, it is based on the idea of enhancing the communication across layer and component boundaries. Application software has a privileged position because they have information about the end user needs of service quality that is not available at other levels. Hence when requesting resources it can specify this information in order to allow more efficient execution of a task. For example, requests for streaming or displaying multimedia can communicate the desired resolution or sampling rate. The latency for network connections might be communicated by the application, allowing for batching of requests or the switch to a more energy efficient transport protocol. An example of the latter is given in [3], where different networking delay requirements of mobile device applications are exploited.

### 6.3.4 Ensemble Optimization

Systems and system parts often do not have a good "energy proportional" behavior. A remedy to this can be to manage multiple instances of these systems or subsystem components together and achieve an energy proportional behavior at the ensemble level. Tolia et al. [34] are an early example that uses an algorithm for automatic virtual machine creation and destruction based on the current and expected workload. Another example for ensemble optimization is based on workload placement; when resources that are not needed by a primary process are used for less time critical tasks. For example, machines that serve web responses during the day can be used for indexing task during the night. A particular example of ensemble optimization can be found in Götz et al. [51]. They developed a framework for dynamic placement of application processes across multiple resources that are optimized as a whole. Part of this setup is the availability of required quality of service information for the applications in order to avoid energy waste due to allocating over-performing resources. One of the big promises of virtualization and especially cloud computing is that it leverages the scale to implement optimizations that cannot be realized on the level of single systems or applications. From the viewpoint of individual application design, it can be a valid strategy to enable ensemble optimization by allowing for virtualization and supporting scaling-out scenarios. The importance of supporting these ensemble level optimizations in the design is, for example, stressed in [50].

### 6.3.5 Automated Code Transformations

There are situations where compilers can automate energy-related optimizations. An example is the compiler designed by Kremer et al. [26] that is able to judge which computations can be offloaded from a mobile device to a server in order to prolong battery life. Of course, many compilers that target mobile platforms already have energy efficiency as a separate optimization goal. But usually the mechanisms do not work at this high level and don't involve networked nodes other than the target device itself.

Another interesting example of using automatic code transformations for building energy efficient programs is the HipHop transformer for PHP [20]. It has been originally developed by Facebook to transform their PHP code into highly efficient C++ code. By being more resource efficient in terms of number of servers required for a certain throughput of web responses, the code is also more energy efficient. Aside from applying automatic code transformation, HipHop can be seen as an instance of energy optimization by plain performance tuning. As was mentioned already, it is important to keep in mind that also for HipHop the main focus has been resource efficiency and not time behavior in terms of response time.

### 6.3.6 Algorithm Optimization

As in the case of performance, there often exist alternative algorithms that provide the desired functionality. Those algorithms often do not have the same energy behavior. There might be an algorithm that is superior to others in terms of every type of resource consumption, but there might also be more subtle differences. For example, one algorithm might be more memory intensive, and another one might be mainly CPU based. Depending on the relative efficiency of these components in the situation at hand, one of them can provide a better overall energy efficiency for a certain computing task. Also the algorithm's energy efficiency might depend greatly on specific characteristics of the workload. Hence, it is not only interesting to look for alternatives which are energy efficient in general, but also to select the best fitting algorithm for the hardware and workload characteristics at hand. An example for application-level algorithm optimization is given in Bunse et al. [10]. They investigate sorting algorithms from the point of view of energy efficiency and develop an algorithm selection routine for a mobile platform.

### 6.3.7 Functional Reduction

The examples given thus far have in common that they are directed towards *computational efficiency*. They take a request for some functionality as given and optimize the computation that delivers it. However, it is often the case that applications do provide functionality that is not really required or not used. Still, it might increase the energy footprint of the application. Think of an increased memory usage because an unused feature is loaded always when the application is started or, of some data displayed to the user that he is not looking at. Scrutinizing the functional and so-called non-functional requirements can be a very effective way of reducing the energy footprints of applications. Application designers have a privileged position compared to engineers that develop lower level software because the former can often influence the requirements side. Also designing applications in a way that functional modules can be loaded and offloaded according to the actual use or according to the user's configuration can improve energy efficiency compared to a monolithic application design.

**Table 3** Optimization strategies mapping on resource consumers and managers.

| Strategy | Address Resource Consumer or Manager? |
|---|---|
| Plain performance optimization | Both |
| Do not stand in the way of power management | Consumer |
| Communicate the required service level | Both |
| Ensemble optimization | Manager |
| Automated code transformations | Consumer |
| Algorithm optimization | Consumer |
| Functional reduction | Consumer |

Table 3 attributes the strategies mentioned to the distinction into resource managers and consumers made in Saxe [31].

# 7. FUTURE CHALLENGES

We have argued that addressing energy efficiency of IT systems at the application software level is very important. Although some results are available, the research area is still considerably young and we expect a substantial growth in attention both from research and from applied software engineering. This section points out some challenges for these activities that will influence the further development in this discipline.

## 7.1 The Playing Field Changes Fast

IT changes at a fast pace and some trends may change the constellation in the near future. Examples are:

- Fast changes in hardware efficiency might change the landscape that optimization heuristics are based on. These trends might drastically reduce the share of specific components of the energy used by the whole system and as a consequence change the appropriateness of optimization depending on the type of application. Consider, for example, the introduction of aggressive power conservation features on all kinds of hardware components. Recently, the CPU has ceased to be the single most energy-consuming component in systems and models that where based on this assumption or at least on the assumption that CPU is a sufficiently accurate proxy for whole system power consumption have to be adjusted. The introduction of solid-state drives has also triggered new strategies of energy optimization. Another factor is the fast improvement to storage and network energy efficiency combined with the general trend of the massive increase of storage and network capacity needs of applications. A bit further down the road are expected developments in memory and disk technology, for example, the memristor [28] that can be quite disruptive for current power management and resource optimization strategies.

- Advances in data center infrastructure potentially diminish the role of server-side energy usage. Together with the rise of the cloud paradigm and other virtualization-based platforms it might increase the importance of enabling black-box ensemble level strategies versus application or system-specific optimizations. Furthermore, the virtualization trends have impact on the ability to instrument and monitor application behavior and the applicability of measurement and modeling approaches.

- Transitions of dominating usage scenarios like the move to mobile computing. These may lead to a growing importance of optimization for the mobile platform and accordingly require optimizations valid for multiple heterogeneous device types and trigger an emphasis on wireless network efficiency.

These changes will influence the design of energy efficient software. It remains important to identify guidelines and best practices for energy efficient system design, but some of the guidelines and the models they are based on may become invalidated. And this may happen at a fast pace. One possible answer to this challenge is an emphasis on tools for modeling, monitoring, and profiling rather than specific design guidelines for developers. It can also be seen as an additional requirement for models and optimization routines to be able to quickly adjust to changes in platform behavior.

## 7.2 Software Outlives Hardware

This is related to the previous point about the dependency of optimization strategies on technological changes. But there is an additional challenge for the designers of software systems. As the guidelines and best practices for energy efficient design change over time it is possible that what has been a best practice for efficient system design some years ago is not valid anymore. In addition to this, the software architect has to cope with the fact that the software system is likely to outlive the hardware and infrastructure it was designed to run on. Use in production for lifespans of more than 10 or even 20 years are not uncommon for software systems. In fact, the platform dependence of old systems has been a major problem for the industry and has stimulated the rise of platform independence and virtual machine technology. This means that systems currently developed are likely to undergo considerable changes in the hardware they are running on. And while the energy efficiency of a system can in principle improve through a closer integration of software with the underlying hardware, deep changes to the latter are conflicting with a tight integration. Fundamental choices taken for good energy behavior of a system may not be valid anymore. And while it may be often possible to adapt the software design to cope with changes, it is by no means guaranteed that this will always be possible or realistic from an economic perspective. Solutions for energy efficiency of the software level will have to have an answer to this fundamental aspect of today's computing ecosystems. A reaction to this challenge could result in different choices in system design concerning the flexibility of the architecture and an increased emphasis on energy monitoring.

## 7.3 Trade-Offs with Other System Properties

As energy efficiency of applications is expected to become even more important, there will be conflicts with other desired system properties that have to be settled by trading one for the other. An obvious candidate that is already covered extensively in the literature is quality of service (QoS) as performance—even though it is not yet systematically understood in which situations a conflict does really occur. Other, maybe less obvious, candidates are reliability, security, maintainability, or functionality. Let's look, for example, at maintainability.

- *Energy efficiency induces increased complexity:* Resource management and optimization routines come at a cost; they consume energy themselves and they increase the overall system complexity and thereby have a negative impact on maintainability. Also, since the software architecture covers an additional design constraint, the design complexity and the danger that basic principles get corrupted during maintenance phase will increase.

- *Abstraction and implementation hiding can reduce efficiency:* Also related to system complexity is the fact that many ways to reduce system complexity for the sake of development productivity and maintainability might have a negative effect on its energy efficiency. There are two main reasons for this. First, abstraction layers, or programming constructs, for example, deep inheritance structures in OO languages, tend to make the execution of a program less efficient. Calls have to be dynamically resolved and routed through these abstraction layers and additional library or framework code will be executed. Second, the encapsulation of library implementation is bound to involve some sort of generality. Library developers have to implement general-purpose solutions that fit a wide range of usage scenarios and cannot be custom tailored to result in the most efficient implementation for a specific context of use. Generality, while supporting reuse, is likely to come with a penalty on resource efficiency in general and energy efficiency in particular [11, 29]. Possible solutions of this problem could involve the extension of library APIs with arguments to specify the required level of service. But it may be also a viable strategy to decide to break the encapsulation of software components on specific points if this can be justified with larger energy savings on the whole application level.

## 8. CONCLUSION

Energy efficiency of IT has received widespread attention, but not yet so in the domain of software engineering. We think it should become a first class system property for software system design and that application-level optimizations should be one of the focus areas of the emerging science of power management. Being a first class system property means that software designers and programmers start to think about energy properties of their systems from the beginning and add power monitoring and modeling facilities to the standard development and production environment. There is also a big non-technical challenge to this; getting energy measurement data from production environments often requires crossing organizational boundaries. Also, energy costs are often not part of the total cost of ownership (TOC) for application owners due to the commonly used billing structure of data centers. This results in a lack of incentives for change. Overall, there is still a lack of knowledge and awareness at the stakeholders' level. Increased visibility of the application energy footprint through measurement, estimation or another form of billing by the data center will play a crucial role in driving changes toward more energy efficient software.

## ACKNOWLEDGMENT

We would like to thank Freek Bomhof from TNO, Netherlands for the discussions and valuable feedback on various versions of this chapter.

## REFERENCES

[1] American Council for an Energy-Efficient Economy (ACEEE), Information and Communication Technologies: The Power of Productivity, How ICT Sectors are Transforming the Economy While Driving Gains in Energy Productivity, Report No. E081, February 2008.

[2] Susanne Albers, Energy-efficient algorithms, Communications of the ACM 53 (5) (2010) 86–96, http://dx.doi.org/10.1145/1735223.1735245.

[3] Niranjan Balasubramanian, Aruna Balasubramanian, Arun Venkataramani, Energy consumption in mobile phones: a measurement study and implications for network applications, in: Proceedings of the 9th ACM SIGCOMM Conference on Internet Measurement Conference (IMC '09), ACM, New York, NY, USA, 2009, 280–293, http://dx.doi.org/10.1145/1644893.1644927.

[4] Luiz André Barroso, Urs Hölzle, The case for energy-proportional computing, Computer 40 (12) (2007) 33–37, http://dx.doi.org/10.1109/MC.2007.443.

[5] Frank Bellosa, The benefits of event-driven energy accounting in power-sensitive systems, in: Proceedings of the 9th Workshop on ACM SIGOPS European Workshop: Beyond the PC: New Challenges for the Operating System (EW 9), ACM, New York, NY, USA, 2000, 37–42, http://doi.acm.org/10.1145/566726.566736.

[6] A. Bianzino et al., A survey of green networking research, IEEE Communication Surveys and Tutorials 99 (2010) 1–18.

[7] J. Blackburn, K. Christensen, A simulation study of a new green bittorrent, in: IEEE International Conference on Communications Workshops 2009, ICC Workshops 2009, 2009, pp. 1–6.

[8] Frédéric Bordage, 24 May 2010. http://www.greenit.fr/article/logiciels/logiciel-la-cle-de-l-obsolescence-programmee-du-materiel-informatique-2748 (An English summary of this comparison of energy consumption of Microsoft Office 97-Office 2010 can be found at: "It's the software, stupid" http://www.guardian.co.uk/sustainable-business/software-energy-efficiency).

[9] David J. Brown, Charles Reams, Toward energy-efficient computing, Communications of the ACM 53 (3) (2010) 50–58, http://dx.doi.org/10.1145/1666420.1666438.

[10] C. Bunse, H. Höpfner, E. Mansour, S. Roychoudhury, Exploring the energy consumption of data sorting algorithms in embedded and mobile environments, in: Proceedings of Mobile Data Management, 2009, pp. 600–607.

[11] Eugenio Capra, Chiara Francalanci, Sandra A. Slaughter, Is software green? Application development environments and energy efficiency in open source applications, Information and Software Technology 54 (1) (2012) 60–71, http://dx.doi.org/10.1016/j.infsof.2011.07.005.

[12] Fay Chang, Keith I. Farkas, Parthasarathy Ranganathan, Energy-driven statistical sampling: detecting software hotspots, in: Babak Falsafi, T.N. Vijaykumar (Eds.), Proceedings of the 2nd International Conference on Power-Aware Computer Systems (PACS'02), Springer-Verlag, Berlin, Heidelberg, 2002, pp. 110–129.

[13] Hui Chen, Weisong Shi, Power measurement and profiling: state-of-the-art, in: Ishfaq Ahmad, Sanjay Ranka (Eds.), Handbook on Energy-Aware and Green Computing, Chapman and Hall/CRC Press, Taylor and Francis Group LLC, 2011.

[14] T. Do, S. Rawshdeh, W. Shi, pTop: a process-level power profiling tool, in: Workshop on Power Aware Computing and Systems (HotPower '09), 2009.

[15] Intel EnergyChecker SDK. <http://software.intel.com/en-us/articles/intel-energy-checker-sdk/>; Intel Energy Checker Software Developer Kit User Guide, Revision 2.0, December 15, 2010.

[16] US Environmental Protection Agency, EPA Report on Server and Data Center Energy Efficiency, August 2007.

[17] Rong Ge, Xizhou Feng, Shuaiwen Song, Hung-Ching Chang, Dong Li, Kirk W. Cameron, PowerPack: energy profiling and analysis of high-performance systems and applications, IEEE Transactions on Parallel and Distributed Systems 21 (5) (2010) 658–671.

[18] Sebastian Götz, Claas Wilke, Sebastian Cech, Uwe Aßmann, Architecture and mechanisms of energy auto-tuning, Sustainable ICTs and Management Systems for Green

Computing, IGI Global, 2012, pp. 45–73, http://dx.doi.org/10.4018/978-1-4666-1839-8.ch003 .

[19] Stavros Harizopoulos, Mehul A. Shah, Justin Meza, Parthasarathy Ranganathan, Energy efficiency: the new holy grail of data management systems research, CIDR, 2009.

[20] HipHop. <https://github.com/facebook/hiphop-php/wiki/>.

[21] ISO/IEC 9126-1:2001 Software Engineering—Product Quality—Part 1: Quality Model. <http://www.iso.org/iso/iso_catalogue/catalogue_tc/catalogue_detail.htm?csnumber=22749>; ISO/IEC 25010:2011: Systems and software engineering—Systems and software Quality Requirements and Evaluation (SQuaRE)—System and software quality models. <http://www.iso.org/iso/iso_catalogue/catalogue_tc/catalogue_detail.htm?csnumber=35733>.

[22] Ravi Jain, David Molnar, Zulfikar Ramzan, Towards understanding algorithmic factors affecting energy consumption: switching complexity, randomness, and preliminary experiments in: Proceedings of the 2005 Joint Workshop on Foundations of Mobile Computing (DIALM-POMC '05), ACM, New York, NY, USA, 2005, pp. 70–79, http://doi.acm.org/10.1145/1080810.1080823.

[23] Aman Kansal, Feng Zhao, Jie Liu, Nupur Kothari, Arka A. Bhattacharya, Virtual machine power metering and provisioning, in: Proceedings of the 1st ACM Symposium on Cloud computing (SoCC '10), ACM, New York, NY, USA, 2010, pp. 39–50, http://dx.doi.org/10.1145/1807128.1807136.

[24] Aman Kansal, Feng Zhao, Fine-grained energy profiling for power-aware application design, SIGMETRICS Performance Evaluation Review 36 (2) (2008) 26-31, http://dx.doi.org/10.1145/1453175.1453180.

[25] Krishna Kant, Toward a science of power management, Computer 42 (9) (2009) 99–101, http://dx.doi.org/10.1109/MC.2009.303.

[26] Ulrich Kremer, Jamey Hicks, James Rehg, A compilation framework for power and energy management on mobile computers, in: Henry G. Dietz (Ed.), Proceedings of the 14th International Conference on Languages and Compilers for Parallel Computing (LCPC'01), Springer-Verlag, Berlin, Heidelberg, 2001, pp. 115–131.

[27] Jorge Luis Zapico, Marko Turpeinen, Nils Brandt, Greenalytics: a tool for mash-up life cycle assessment of websites, in: Proceedings of the 24th International Conference on Informatics for Environmental Protection, Shaker Verlag, Aachen, Germany, 2010.

[28] The memristor. <http://en.wikipedia.org/wiki/Memristor>.

[29] Parthasarathy Ranganathan, Recipe for efficiency: principles of power-aware computing, Communications of the ACM 53 (4) (2010) 60–67, http://dx.doi.org/10.1145/1721654.1721673.

[30] Suzanne Rivoire, Parthasarathy Ranganathan, Christos Kozyrakis, A comparison of high-level full-system power models, in: Proceedings of the 2008 Conference on Power Aware Computing and Systems (HotPower'08), USENIX Association, Berkeley, CA, USA, 2008, p. 3.

[31] Eric Saxe, Power-efficient software, Communications of the ACM 53 (2) (2010) 44–48, http://dx.doi.org/10.1145/1646353.1646370.

[32] Chiyoung Seo, George Edwards, Daniel Popescu, Sam Malek, Nenad Medvidovic, A framework for estimating the energy consumption induced by a distributed system's architectural style, in: Proceedings of the 8th International Workshop on Specification and Verification of Component-Based Systems (SAVCBS '09), ACM, New York, NY, USA, 2009, pp. 27–34, http://dx.doi.org/10.1145/1596486.1596493.

[33] Dimitris Tsirogiannis, Stavros Harizopoulos, Mehul A. Shah, Analyzing the energy efficiency of a database server, in: Proceedings of the 2010 ACM SIGMOD International Conference on Management of Data (SIGMOD '10), ACM, New York, NY, USA, 2010, pp. 231–242, http://dx.doi.org/10.1145/1807167.1807194.

[34] Niraj Tolia, Zhikui Wang, Manish Marwah, Cullen Bash, Parthasarathy Ranganathan, Xiaoyun Zhu, Delivering energy proportionality with non energy-proportional

systems: optimizing the ensemble, in: Proceedings of the 2008 Conference on Power Aware Computing and Systems (HotPower'08), USENIX Association, Berkeley, CA, USA, 2008, p. 2.

[35] D. Tsirogiannis, S. Harizopoulos, M.A. Shah, Analyzing the energy efficiency of a database server, in: SIGMOD '10: Proceedings of the 2010 International Conference on Management of Data, ACM, New York, NY, USA, 2010, pp. 231–242.

[36] Shinan Wang, Hui Chen, WeiSong Shi, SPAN: a software power analyzer for multicore computer systems, Sustainable Computing: Informatics and Systems 1 (1) (2011) 23–34, ISSN 2210-5379, 10.1016/j.suscom.2010.10.002, <http://www.sciencedirect. com/science/article/pii/s221053791000003x>.

[37] Hewlett-Packard Corporation, Intel Corporation, Microsoft Corporation, Phoenix Technologies Ltd., Toshiba Corporation. ACPI, Advanced Configuration and Power Interface, specification available online, Revision 4.0, June 2009. <http://www.acpi. info/spec.htm>.

[38] Stefan Naumann, Markus Dick, Eva Kern, Timo Johann, The GREENSOFT model: a reference model for green and sustainable software and its engineering, Sustainable Computing: Informatics and System 1 (4) (2011) 294–304, http://dx.doi.org/10.1016/ j.suscom.2011.06.004, ISSN: 2210-5379. <http://www.sciencedirect.com/science/ article/pii/S2210537911000473>.

[39] Alain J. Martin, Towards an energy complexity of computation, Information Processing Letter 77 (2–4) (2001) 181–187, http://dx.doi.org/10.1016/S0020-0190(00)00214-3.

[40] Joulemeter. <http://research.microsoft.com/en-us/projects/joulemeter/default.aspx>.

[41] pTop. <http://mist.cs.wayne.edu/ptop.html>.

[42] Greenalytics. <http://www.greenalytics.org/>.

[43] Green Certified Site. <http://www.co2stats.com/>.

[44] Google Analytics. <http://www.google.com/analytics/>.

[45] Niklaus Wirth, A plea for lean software, Computer 28 (2) (1995) 64–68, http://dx.doi. org/10.1109/2.348001.

[46] W.S. Jevons, The Coal Question: An Inquiry Concerning the Progress of the Nation, and the Probable Exhaustion of our Coal-Mines, second ed., Macmillan and Co, London, 1865, 1866.

[47] Kien Le, Ricardo Bianchini, Thu D. Nguyen, Ozlem Bilgir, Margaret Martonosi, Capping the brown energy consumption of Internet services at low cost, in: Proceedings of the International Conference on Green Computing (GREENCOMP '10), IEEE Computer Society, Washington, DC, USA, 2010, pp. 3–14, http://dx.doi.org/10.1109/ GREENCOMP.2010.5598305.

[48] LessWatts.org. <http://www.lesswatts.org/projects/applications-power-management/ avoid-pulling.php.

[49] PowerTOP. Official Website. <https://01.org/powertop/>.

[50] M. Aggar, The IT Energy Efficiency Imperative, Whitepaper from Microsoft Corporation, June 2011. <http://download.microsoft.com/download/7/5/A/75AB83E8- 2487-409F-AC6C-4C3D22B72139/ITEI_Paper_5.27.11.pdf>.

[51] Sebastian Götz, Claas Wilke, M. Schmidt, Sebastian Cech, Uwe Aßmann, Towards energy auto tuning, in: Proceedings of 1st Annual International Conference on Green Information Technology – GREEN IT, 2010.

[52] Narendran Thiagarajan, Gaurav Aggarwal, Angela Nicoara, Dan Boneh, Jatinder Pal Singh, Who killed my battery?: analyzing mobile browser energy consumption, in: Proceedings of the 21st International Conference on World Wide Web (WWW '12), ACM, New York, NY, USA, 2012, pp. 41–50, http://dx.doi.org/10.1145/2187836.2187843.

## ABOUT THE AUTHORS

**Kay Grosskop** studied philosophy at the University of Amsterdam, The Netherlands. He currently works as a researcher and software engineer at the Software Improvement Group (SIG) in Amsterdam.

**Joost Visser** carried out his PhD research at the Centre for Mathematics and Computer Science (CWI) in Amsterdam. He was a researcher in the LMF Group at the Departamento de Informática of the Universidade do Minho in Braga, Portugal, and member of the Computer Science and Technology Center (CCTC). Joost is currently Head of Research at the Software Improvement Group (SIG) in Amsterdam, The Netherlands. He also holds a part-time position as Professor of "Large-Scale Software Systems" at the Radboud University Nijmegen, The Netherlands.

# AUTHOR INDEX

# SUBJECT INDEX

*Note*: Page numbers followed by "*f*" and "*t*" indicate figures and tables respectively

# CONTENTS OF VOLUMES IN THIS SERIES

## Volume 71

Programming Nanotechnology: Learning from Nature
    BOONSERM KAEWKAMNERDPONG, PETER J. BENTLEY, AND NAVNEET BHALLA
Nanobiotechnology: An Engineers Foray into Biology
    YI ZHAO AND XIN ZHANG
Toward Nanometer-Scale Sensing Systems: Natural and Artificial Noses as Models for
Ultra-Small, Ultra-Dense Sensing Systems
    BRIGITTE M. ROLFE
Simulation of Nanoscale Electronic Systems
    UMBERTO RAVAIOLI
Identifying Nanotechnology in Society
    CHARLES TAHAN
The Convergence of Nanotechnology, Policy, and Ethics
    ERIK FISHER

## Volume 72

DARPAs HPCS Program: History, Models, Tools, Languages
    JACK DONGARRA, ROBERT GRAYBILL, WILLIAM HARROD, ROBERT LUCAS,
    EWING LUSK, PIOTR LUSZCZEK, JANICE MCMAHON, ALLAN SNAVELY, JEFFERY VETTER,
    KATHERINE YELICK, SADAF ALAM, ROY CAMPBELL, LAURA CARRINGTON,
    TZU-YI CHEN, OMID KHALILI, JEREMY MEREDITH, AND MUSTAFA TIKIR
Productivity in High-Performance Computing
    THOMAS STERLING AND CHIRAG DEKATE
Performance Prediction and Ranking of Supercomputers
    TZU-YI CHEN, OMID KHALILI, ROY L. CAMPBELL, JR., LAURA CARRINGTON,
    MUSTAFA M. TIKIR, AND ALLAN SNAVELY
Sampled Processor Simulation: A Survey
    LIEVEN EECKHOUT
Distributed Sparse Matrices for Very High Level Languages
    JOHN R. GILBERT, STEVE REINHARDT, AND VIRAL B. SHAH
Bibliographic Snapshots of High-Performance/High-Productivity Computing
    MYRON GINSBERG

## Volume 73

History of Computers, Electronic Commerce, and Agile Methods
    DAVID F. RICO, HASAN H. SAYANI, AND RALPH F. FIELD
Testing with Software Designs
    ALIREZA MAHDIAN AND ANNELIESE A. ANDREWS
Balancing Transparency, Efficiency, and Security in Pervasive Systems
    MARK WENSTROM, ELOISA BENTIVEGNA, AND ALI R. HURSON
Computing with RFID: Drivers, Technology and Implications
    GEORGE ROUSSOS

## Volume 79

## Volume 80

## Volume 81

**Volume 87**

Printed and bound by CPI Group (UK) Ltd, Croydon, CR0 4YY

03/10/2024

01040425-0005